Using Commercial Contracts

Using Commercial Contracts:

A practical guide for engineers and project managers

David Wright

Consultant and lecturer on commercial contracts
Visiting lecturer at Manchester University
Sometime visiting lecturer and fellow at Cranfield University

This edition first published 2016
© 2016 by John Wiley & Sons Ltd

Registered office
John Wiley & Sons Ltd, The Atrium, Southern Gate, Chichester, West Sussex, PO19 8SQ, United Kingdom.

Editorial offices
9600 Garsington Road, Oxford, OX4 2DQ, United Kingdom.
The Atrium, Southern Gate, Chichester, West Sussex, PO19 8SQ, United Kingdom.

For details of our global editorial offices, for customer services and for information about how to apply for permission to reuse the copyright material in this book, please see our website at www.wiley.com/wiley-blackwell.

The right of the author to be identified as the author of this work has been asserted in accordance with the UK Copyright, Designs and Patents Act 1988.

British Library Cataloguing-in-Publication Data

A catalogue record for this book is available from the British Library

Library of Congress Cataloging-in-Publication Data

Names: Wright, David, 1939 November 11- author.
Title: Using construction contracts : a practical guide for engineers and
   project managers / David Wright.
Description: Chichester, West Sussex, United Kingdom ; Hoboken : John Wiley &
   Sons, Ltd, 2017. | Includes index.
Identifiers: LCCN 2016003746| ISBN 9781119152507 (pbk.) | ISBN 9781119152545
   (epub)
Subjects: LCSH: Construction contracts–England. | Construction
   contracts–Wales.
Classification: LCC KD1615 .W75 2017 | DDC 343.4207/8624–dc23 LC record available at
   http://lccn.loc.gov/2016003746

Wiley also publishes its books in a variety of electronic formats. Some content that appears in print may not be available in electronic books.

Cover image: GettyImages/tmeks

Set in 10/13pt HelveticaNeue by SPi Global, Chennai, India.
Printed and bound in Singapore by Markono Print Media Pte Ltd

1 2016

Contents

About the Author

David Wright left university with a good law degree. He worked as a commercial lawyer/manager/negotiator in the electrical industry, then, successively in the electronics/avionic and defence industries, on the client side of the offshore-oil industry, and the contracting side of the chemical/process engineering industries. Finally he spent three years at board level, in charge of the contractual and commercial function, in a joint venture process contractor, followed by a period as General Counsel in an electrical/mechanical engineering group of companies.

Since then he has been a lecturer and consultant on contracts and related areas in the electrical/mechanical and process engineering industries. He was a member for over seventeen years of the committees of the Institution of Chemical Engineers which write the Institution's model conditions of contract for the process industry, and supervise the approval, selection and training of arbitrators and adjudicators. He has taught at a number of universities within the UK in law, contract conditions, negotiation and contract/project management, including at Cranfield University, where he held successively a visiting lectureship and visiting fellowship in European Business Law, and at the University of Leeds, Imperial College, and Manchester University, where he has held a visiting lectureship for some twenty years.

He is the author of the *Purple Book*, the standard Guide to the ICHEME model conditions of contract, *A Guide to Consultancy* and *Law for Project Managers*. Finally, he is an experienced arbitrator, mediator and adjudicator in the process industry.

Preface

Many years ago I was in Moscow, carrying out a negotiation in the offices of a state buying organisation. After my negotiation was finished I had to wait for the formal signatures to the document that I had agreed. I passed the time sitting in one of the negotiation rooms. While I was waiting, a Russian buyer, whom I knew, came into the room to discuss a contract with a Japanese salesman. The contract was for the purchase of spare parts for some Japanese equipment installed in a Russian plant. It was written in English, and the discussion was taking place in English. It was the only language the two negotiators had in common. But there was a problem. As soon as the discussion got into detail the negotiators could not understand each other. The Russian buyer had been taught a completely different style of English and vocabulary to the Japanese salesman. This, combined with differences in pronunciation, made it virtually impossible for them to communicate with each other.

I, on the other hand, could understand both of them without any difficulty. Finally I intervened, and spent an interesting and very instructive hour as an interpreter from 'Russian English' into 'Japanese English' and back again.

The last point that arose during the negotiation was the law that was to govern the contract. The draft that was being discussed was 'subject to Swiss law'. However, one of the parties, I cannot remember which, objected. As a result it was agreed that the contract should be 'subject to English law', which resulted in a few small standard changes to the text.

This is a nice little story. However, it illustrates three points.

The first is the basic problem of the English language. As a general commercial language English is perfect. It is easy to use, because it is a flexible and slow-spoken language. A basic vocabulary and a limited command of grammar will take you a long way in normal commercial discussions. However, people can speak different kinds of English and speak it in different ways, and not understand other kinds.

The second is that in some senses law is not very important. Provided that the contract clearly spells out what the obligations of each of the parties are, and the parties understand what the contract says, then the 'law of the contract' may be of only minor importance. Swiss law or English law may make little practical difference to a complex contract once everything else has been agreed. It will usually only be necessary to make minor adjustments to a few clauses to make the change. Of course the 'law of the contract' may make a considerable difference if the contract goes badly or a formal dispute arises later on. But the vast majority of negotiated contracts do not result in disputes – even in the civil engineering industry.

The third is that only a little help is necessary to bridge the gap between people who don't understand each other.

This book is the result of years of practice in talking about the law of contract to non-lawyers. It is also the result of what I have learned from others. I have been lucky enough to have worked with some of the best. My tutors at university were A W Brian Simpson and Robert Goff, who tried to get at least some understanding of how law works into my thick head. In the chemical engineering industry I worked for John Brandler, the MD of Petrocarbon Developments Ltd, perhaps the wisest man I have ever known. I owe a debt to Garth Ward and the late Ralph Levene at Cranfield, and to Peter Thompson and Stephen Wearne at Manchester University, and also to all my colleagues on the Institution of Chemical Engineers' committees on dispute resolution and conditions of contract. To these and all the many other people that I have worked with I owe my thanks.

Notes

As you will see, we have tried to explain and illustrate the law by referring to actual cases when it is practicable to do so. When one does this, there is a problem. Many of the basic principles of the law were settled long ago, not in the twenty-first or even in the twentieth century, and there is often a whole series of cases to choose from. We have tried to select cases which give clear examples of the issues, rather than simply selecting the first or the latest.

Women make very good lawyers, and very good negotiators. The author has been skinned alive by women negotiators in what used to be Eastern Europe more often than he cares to admit. However, when writing a book one has to choose a gender. So, with apologies, the text uses 'it' to denote legal persons and 'he' to denote real persons, except where the real person concerned is a she.

Foreword

> This bond doth give thee here no jot of blood;
> The words expressly are 'a pound of flesh':
> Then take thy bond, take thou thy pound of flesh;
> Shed thou no blood; nor cut thou less, nor more,
> But just a pound of flesh; if thou tak'st more
> Or less, than a just pound, be it but so much
> As makes it light or heavy in the substance,
> Or the division of the twentieth part
> Of one poor scruple, nay, if the scale do turn
> But in the estimation of a hair,
> Thou diest and all thy goods are confiscate.
>
> **The Merchant of Venice – Act IV Scene 1 – lines 307–9 and 326–33**

Shakespeare knew the world of contracts. As the owner–manager of an important commercial theatre he needed to. So in *The Merchant of Venice* he puts a commercial dispute on stage. The contract is between a merchant, whose ships have been lost, and a moneylender – the modern equivalent might be a company with a severe liquidity problem and an overdraft. Everyone – judge, claimant and defendant – knew what the contract said, that is until a smarty-pants lawyer stood up and pointed out exactly what those precise words meant. At that point the claimant's case fell apart because it was impossible to perform.

Great theatre – but was the lawyer correct? He, or, as the audience knew, she, was actually making two points. 'Flesh' does not mean or include blood, and a pound means *precisely* a pound.

But was it the claimant's problem if the defendant lost blood as a result of the contract being carried out? Wasn't that actually the defendant's risk? (And Shylock had already said 'If you prick us do we not bleed?') Also *de minimis non curat lex*, the law is not bothered about trifling inaccuracies.

So what was Shakespeare saying?

Perhaps a basic working knowledge of law may be useful. A clever lawyer can run rings round a commercial manager. The lawyer may often get away with it even if the legal arguments are a bit suspect. And, finally, perhaps when it comes to sensitivity to the precise meaning of language, women are usually more skilful than men.

Judge for yourselves.

(And by the way, if anyone doubts this last statement, a sizeable study some years ago by a major UK university found it to be completely correct.)

Cases Referred to

Adams *v*. Lindsell (1818) 1 B & Ald 681 – Chapter 5

AEG (UK) Ltd *v*. Logic Resources Ltd [1996] CLC 265 – Chapter 10

Ailsa Craig Fishing Co Ltd *v*. Malvern Fishing Co Ltd [1983] 1 WLR 964; 1 All ER 101 – Chapter 10

Allcard *v*. Skinner (1887) 36 Ch D 145 – Chapter 11

Allied Marine Transport Ltd *v*. Vale do Rio Doce Navegacao SA [1984] 1 WLR 1; [1983] 3 All ER 737 – Chapter 5

Aluminium Industrie Vassen B V *v*. Romalpa Aluminium Ltd [1976] 1 WLR 676; 2 All ER 552 – Chapter 17

Amalgamated Investments and Property Co Ltd *v*. Texas Commercial Int. Bank [1982] QB 84 – Chapter 17

Ampurius Nu Homes Holdings Ltd *v*. Telford Homes (Creekside) Ltd [2013] 4 All ER 377 – Chapter 9

Anderson Ltd *v*. Daniel [1924] 1 KB 138 – Chapter 12

Andrews Bros (Bournemouth) *v*. Singer & Co Ltd [1934] 1 KB 17 – Chapter 10

Archbolds (Freighterage) Ltd *v*. S Spanglett Ltd [1961] 1 QB 374; 2 WLR 170; 1 All ER 417 – Chapter 12

Arcos Ltd *v*. E A Ronaason & Son [1933] AC 470 – Chapters 8 and 9

Ashington Piggeries *v*. Christopher Hill Ltd [1972] AC 441; 1 All ER 847 – Chapter 9

Ashmore Benson Pease & Co Ltd *v*. A V DawsonLtd [1973] 1 WLR 828; 2 All ER 856 – Chapter 12

Associated Japanese Bank (International) Ltd *v*. Credit du Nord SA [1989] 1 WLR 255; (1988) 3 All ER 902 – Chapters 9 and 11

Atlas Express Ltd *v*. Kafco (Importers and Distributors) Ltd [1989] 1 QB 833; 3 WLR 389; 1 All ER 64 – Chapter 11

Attwood *v*. Small (1838) 6 Cl & F 232 – Chapter 6

B & S Contracts and Design Ltd *v*. Victor Green Publications Ltd (1984) 128 SJ 279 – Chapter 11

Bainbridge *v*. Firmston (1838) 8 A&E 743 – Chapter 4

Balfour *v*. Balfour [1919] 2 KB 571 – Chapter 4

Bank of Credit and Commerce International SA *v*. Aboody [1990] 1 QB 923; [1989] 2 WLR 759; [1992] 4 All ER 955 – Chapter 11

Bannerman *v*. White (1861) 10 CB (NS) 844 – Chapters 6 and 7

Barclays Bank plc *v*. O'Brien [1994] 1 AC 180; [1993] 3 WLR 786; [1993] 4 All ER 417 – Chapter 11

Barrow Lane & Ballard Ltd *v*. Philip Philips & Co Ltd [1929] 1 KB 574 – Chapter 11

Statutes Referred to

1870	Apportionment Act
1874	Infants Relief Act
1882	Married Women's Property Act
1882	Bills of Exchange Act
1890	Partnership Act
1893	Sale of Goods Act
1907	Limited Partnership Act
1925	Law of Property Act
1925	Trustee Act
1933	Pharmacy and Poisons Act
1943	Law Reform (Frustrated Contracts) Act
1954	Protection of Birds Act
1957	Occupiers Liability Act
1959	Restriction of Offensive Weapons Act
1959	Mental Health Act
1961	Housing Act
1964/65	Hire Purchase Acts
1967	Misrepresentation Act
1971	Powers of Attorney Act
1972	European Communities Act
1973	Supply of Goods (Implied Terms) Act
1973	Fair Trading Act
1974	Consumer Credit Act
1974	Health and Safety at Work Act
1974	Solicitors Act
1976	Resale Prices Act
1976	Restrictive Trade Practices Act
1977	Torts (Interference with Goods) Act
1977	Unfair Contract Terms Act
1979	Sale of Goods Act
1980	Competition Act
1980	Limitation Act
1981	Supreme Court Act
1982	Supply of Goods and Services Act
1983	Mental Health Act
1985	Companies Act
1985	Enduring Powers of Attorney Act
1986	Building Societies Act

1

The Law in General

This book is about commercial, as opposed to consumer, contract law. It does not deal with questions of consumer protection and consumer rights. So it makes no reference to such matters as consumer legislation (which has culminated in the Consumer Rights Act 2015, coming into force at the time of writing and which promises to be of great importance), except where it affects contract law generally.

The book is about the law of England and Wales, though in practice the contract laws of the four other legal systems within the United Kingdom will not be much different. (They are the legal systems of Scotland, Northern Ireland, the Isle of Man and the Channel Islands.)

Law is based around the principle of the nation-state. Each country has its own legal system and its own laws. So a dispute under a contract subject to French law will be decided by the procedures and principles that govern French law, a contract subject to German law by German procedures and principles, and so on. An attempt a few years ago to combine sharia law with English law in a contract, *Shamil Bank of Bahrain EC* v. *Beximco Pharmaceuticals Ltd* (2004), was rejected. The court stated that one cannot combine legal systems. (A judge under any other system would say the same.) However, the court did accept that it might well be possible for a contract governed by English law to incorporate a specific rule from another legal system.

A contract between two English companies carried out in England will be subject to English law, almost automatically, and any dispute will be settled in the UK. But with any trans-national contract there will be a choice. The choice decides which law will govern the contract and apply to any dispute, and perhaps also the place where that dispute will be decided. With most complex commercial contracts the broad outlines of law will be much the same in most

Using Commercial Contracts: a practical guide for engineers and project managers, First Edition. David Wright.
© 2016 John Wiley & Sons, Ltd. Published 2016 by John Wiley & Sons, Ltd.

countries. However, the detail may well be very different, and the dispute procedures will be different. The parties should make a conscious choice, not an accidental one. See the case of *Entores* v. *Miles Far East Corporation* (1955) in Chapter 4 below.

The point is covered by the Contracts (Applicable Law) Act 1990, and the Private International Law (Miscellaneous Provisions) Act 1995, with reference to Article 4(2) of the 1980 Rome Convention on the Law Applicable to Contractual Obligations. These provide that the law applicable to any contract should be governed by the 'principal place of business' of the parties, or one of them, unless under the terms of the contract performance was to be elsewhere, when another law might be selected. It was discussed in the case of *Enstone Building Products Ltd* v. *Stanger Ltd* (2004) which concerned a contract between two English companies for an investigation of building materials to be carried out in Scotland, with a report then to be submitted by the English office of one party to the other. No applicable law was stated in the contract. The court decided that the contract was governed by English law.

There are a number of preliminary points to remember.

1.1 A little understanding

Almost everyone can drive a car. Very few of us are skilled mechanics or physicists. We do not understand Newton's first and second laws of motion or how to take a car engine apart. We do not need to. But an understanding of the basics does help if the road is icy or the car has a mechanical problem. Contracts are just the same. Most of the time we do not have any difficulties, but the knowledge is useful, just in case.

1.2 Why contract law?

We live in a market economy. Almost every country in the world is a market economy, more or less. The economic theory behind the market economy is the maximisation of special skills. Everyone should concentrate upon his own special skills, then sell the results of those special skills and buy the results of other people's skills. In that way everyone can maximise his own output, and so the economy as a whole can prosper.

The contract is the basic tool of the market economy. It creates certainty that, barring accidents, a seller will deliver and a buyer will pay. It simply needs to have a clear set of rules and be supported by an efficient enforcement/dispute resolution system such as that of the courts or arbitration. The aim of the law of contract is to provide the rules.

Because the system needs contracts the process of making and managing the contract must not be too difficult. If it becomes too difficult, bearing in mind that the vast majority of contracts are either made by consumers who know little if anything of the niceties of the law or commercial people who are too busy to give more than the minimum time to the exercise, then the system will not work. So it is essential that the rules should be simple, and if they cannot be simple then at the least they must be fairly straightforward and easy to apply to the vast majority of real-life situations.

1.3 Commerce and judges

There is a significant difference between the judge and the organisation. The organisation must react immediately to a situation, or as nearly immediately as makes no difference. Sometimes it has to react without full information. A judge has the luxury of looking at a dispute with time at his disposal and with all the benefits of hindsight. Anyone reading the report of a commercial dispute that has gone to court over the past 30 to 40 years will realise that the events which gave rise to the litigation had happened usually at best some three or four years before the court hearing, and often much earlier than that.

The commercial organisation needs certainty. It needs to know precisely what obligations it has under the contract, and what powers it has to manage the contract. Every legal adviser will find himself faced with different varieties of the same question, on numberless occasions. 'I have a contract; it has gone wrong; what can/should/must I do in these circumstances?' This then leads to two further questions. 'Am I in breach, or is he in breach, or are both of us in breach? Secondly, if he is in breach, am I entitled to a significant sum in damages, and, if I decide that I want to do so, can I terminate?' The organisation needs to know the answers to these questions probably on the day the contract is signed, and certainly on the day when a breach occurs, because it must act correctly. If it does not, it may either lose a legitimate opportunity or risk being dragged into court by the other side.

In contrast to this a judge wants flexibility. When he finds himself faced with a dispute in court, one party will claim that the contract justified him in what he did and the other side will claim the opposite. The function of the judge is give a decision which is legally correct, but which will also produce a fair result in all the circumstances, based upon the evidence that the parties put before him. If all he can do is treat the contract as a rigid book of rules, then his ability to be fair is limited.

1.4 Dispute resolution

Many years ago almost every dispute ended in litigation, before a judge. Now that is not so. The great majority of disputes go elsewhere – to arbitration, adjudication, or, increasingly, mediation/conciliation, all of which are very much quicker. The advantage of litigation is a guarantee of the legal competence/quality of the tribunal. But its twin problems are time and cost.

1.5 Where does the law come from?

Contract law is what is called a 'common law' area of law. This means that most of the law that we have today was not created by parliament, but by the decisions of judges presiding over cases in the courts (over a period of 150 years or more). This has serious implications for the way the law works in practice.

It means that there is often no easy way for the layman to find out what the latest position is, first, because there is no easy point of access, second, because judges often discuss cases, not principles, and third, because the various law reports cover hundreds of thousands of cases. It can be equally difficult for the professional lawyer.

It means that rules can remain in place simply because of accident rather than because of intelligent design.

It also means that the development of the law is often haphazard. It depends to an unhealthy extent upon the cases that have actually come before the courts. Inevitably those cases come from the 'difficult' areas. No one can fail to be struck by the high proportion of cases originating in consumer protection, the shipping world, local government, building/civil engineering, liability insurance and the second-hand car trade. In contrast there are very few cases about disputes in the service, oil, electrical, mechanical, chemical/processing engineering, electronics or IT industries.

What is more, 'normal' disputes seldom come to court – they are settled by negotiation, mediation/conciliation, or perhaps arbitration or adjudication. If occasionally they do go to litigation they are usually settled before the court stage is reached. Many of the cases that come to court involve non-normal circumstances – fraud, sharp practice or total incompetence by one of the parties, a major shift in market values or a major accident or loss. This creates a difficulty for the judge – does he apply the law with cold, impartial objectivity, or does he try to give some degree of justice to the injured party? Judges are human beings like everyone else. The problem is that when judges try to be 'fair', they sometimes bend the law or facts a little. This affects the way the law is explained in judgments, with such judgments then creating the law.

1.6 Keeping law up to date

The common law precedent system has served the UK brilliantly, in the sense that judges of high quality have made decisions that have created a system of contract law that in general fits the needs of society.

But it does have one serious, if occasional, disadvantage. It cannot easily update itself. Every system needs to be updated, and law is no exception. Some decisions will be wrong. Some turn out to have unforeseen consequences. A decision that was completely adequate for the world of 1900 may be hopelessly inadequate in the world of 2000. What are the alternatives? First, someone can bring a test case and then appeal it up to the appropriate level so that the previous decision can be reversed (cancelled out), or qualified (changed or adapted). This is expensive in time and resources. It happens only rarely. Second, which happens occasionally, is that a particularly unfortunate decision will be 'distinguished on the facts'. Whenever it is raised the court will say that the facts in the case now before it are different. As the facts are different, the law can be different. Finally, law can be, and most usually is, updated by legislation. The difficulty here is that this can also mean waiting for several years until parliament can find an empty slot in a packed legislative schedule.

1.7 Objectivity and subjectivity

There are two approaches that one can take to motives. One is to consider what the person actually thinks – a subjective approach. The other is to consider what a reasonable, rational and intelligent person in that position might or should have thought – an objective approach. Over the years, lawyers, both academic and others, have argued at length about which approach the law should take.

The law assumes that in the commercial world the parties to the contract are more or less equal in understanding, and are also professionals – so that they behave in a rational way. Therefore all that they need is to know is what the rules are, and once those rules are clear, they can play the game. So the law for commercial contracts always aims to be objective, taking little account of what individual motives are, except where some degree of deceit or fraud is involved. In the world of consumer contracts the law takes a much more subjective approach.

1.8 Is law 'fair'?

Well, sometimes. The original principles of the law as created in the Victorian age were that every person was competent and could look after his best interests. But even then the law recognised that the vulnerable, children for example, needed some protection. In the modern era the concept of fairness has extended into areas such as consumer and employment protection and the control of unbalanced competition and unfair contract terms, particularly in relation to liability limitation and exclusion clauses.

But in some areas the law does not aim to be fair. In particular, in the area with which this book is mostly concerned, commercial contracts, the law expects competent people to be able to look after themselves. If they cannot do so (see *L'Estrange* v. *Graucob* (1934) below as a prime example), the law will not protect them. Another area where the law is not 'fair' is in litigation or arbitration. Here the law is 'just', simply giving each side the opportunity to present its case and contest the case put forward by the other side. Then the party that puts forward the better case (or puts its case in a better or more convincing way) wins.

1.9 But there is a further question – does law require *the parties* to be fair?

The answer to this question is worth a complete book on its own. There are conflicting principles.

- The law expects competent people to be professional. If I am rather more professional than you, or have rather more bargaining power than you, the contract may favour me more than it favours you. That is only to be expected.
- Second, as has been said, all is fair when in a dispute.
- Third, the law expects the parties to comply more or less exactly with what they have agreed to do in their contract. And the law expects the contract to set out all the rules that are to apply. As long as the parties comply, the law does not require them to be fair in the way in which they comply.

But –

- The law will do its best not to allow a party to a contract to benefit where there is clear evidence of sharp practice. The law on misrepresentation and unilateral mistake are examples of this.
- The law imposes some standards of behaviour on the parties, such as terms that are implied by law or legislation – restrictions on 'unfair' clauses and on clauses in unfair restraint of trade, or restrictions under the Unfair Contract Terms Act or the provisions of the sale of goods legislation and other legislation for reasonable levels of quality and so on, or the doctrine of promissory estoppel.
- The law expects certain standards of conduct in various types of relationship, such as employment or agency contracts.

- There is a solid body of legislation and practice supporting the rights of the consumer to fair treatment.
- Also, in respect of contracts made by private people as opposed to commercial organisations, the law takes at times a rather more subjective approach.

1.10 The two aspects of commercial contracts

Finally, remember that there are two aspects to many commercial contracts.

There is the *transaction*, perhaps the purchase or sale. It is feasible to set out all rules governing the transaction in a contract.

But in addition there is the *relationship*, maybe a joint venture to develop a market or product or a frame contract to cover long-term supply/procurement through a supply chain. These contracts may need to last for a long time and to change to meet unforeseen and unforeseeable circumstances. Here it is simply not always practicable to set out all the terms in the contract.

There has been a steady change in a number of industries towards working within the context of longer-term relationships. This trend is slowly being recognised by law; see for example the 2013 case of *Yam Seng Pte Ltd* v. *International Trade Corporation* where the court was prepared to put a significant amount of emphasis on the need for the parties within a contract relationship to treat each other fairly.

1.11 What does contract law set out to do?

The question is a simple one. The answer is not.

First, contracts can be of many different sizes and types and be intended to operate in many different types of situation. So there is an exception to almost every rule. What is a good rule for auction sales will not fit the world of standard commercial purchasing, for example.

Then the law sets out to do several completely different things at the same time. Sometimes these will be inconsistent, and sometimes they will be directly in conflict with each other.

1. The law is there to provide a framework within which the competent individual or organisation can operate.

   In this sense contract law is just like the law that requires people in the UK to drive on the left, or people in France to drive on the right. It sets out clear and sensible rules that people can follow easily, and so use the law to get on with their business.

2. The law is designed to allow the maximum possible freedom to people and organisations to make whatever contracts they wish and to manage those contracts in accordance with normal practice.

   The idea of freedom of contract is to some extent a historical accident. It comes from the fact that most of the law was worked out during the laissez-faire non-interventionist Victorian age, but people in general and commercial organisations in particular know their business better than the law. So the law does try not to interfere with normal contractual arrangements more than the absolute minimum.

3. The law is intended to protect the private person against the power of the organisation to treat him unfairly.

   The concepts of consumer protection and employee protection have a considerable impact upon various aspects of contract law, especially where badly drafted or 'unfair' clauses limiting or excluding one party's liability to the other for breach of contract are concerned.

4. The law in a modern state must regulate commercial behaviour in the national interest.

   Increasingly the law sets out to regulate various aspects of commercial activity in the overall interests of the public/economy at large. This leads to an interventionist approach by government, whatever its colour. There is and will always be a constant stream of legislation to regulate various aspects of commercial contracts.

5. The law is there to prevent the unrestricted use of monopolistic power to distort normal commercial behaviour.

   Whilst this is largely only peripheral to this book, any organisation dealing with matters such as contracts for the licensing of intellectual property or the creation of a distributor network within the European Union will need to be well aware of the regulatory requirements.

6. The law sets out to provide a framework within which disputes can be settled by 'legal' means.

   The law never likes to tie the hands of the courts too tightly. It will always therefore allow a considerable degree of flexibility to the dispute tribunal to allow it to make a 'fair' decision. So it will always allow the lawyers to 'do a Portia' and play with the words of the contract, if the parties give them the opportunity.

The overall result is that the law is straightforward in principle, but can also be anything but straightforward when applied to a complex situation.

1.12 Transaction versus relationship

The law of contract is not really designed to deal with relationships. The basic theory is that the contract must include all the rules that are to govern the way that the contract is to be carried out, which is completely acceptable in a contract to buy a car, but not really practicable in a contract to collaborate in a long-term project or trading relationship. It has developed to a large extent in connection with transactions – sale, purchase, repair, transport and so on. So, in a transaction contract dispute we can usually predict what a judge will decide.

But the modern world is more and more concerned with long-term relationships, between customers and regular suppliers, the various partners involved in long-term projects and so on. The courts see very few disputes about relationships (and of these the most common are employment and agency contracts). So it is often going to be difficult to predict what a judge might decide in these contexts.

2

The Start – Using an Agent

In law every person, private or commercial, individual or corporate, makes his/its own contracts. In reality we often use agents to make contracts for us. Even in private life we use agents from time to time, estate agents and solicitors to sell or buy our houses, letting agents when we rent, travel agents when we go on holiday, websites to buy concert, coach or rail tickets or to book hotel rooms, and so on. In the commercial world every organisation uses agents to make and manage contracts almost all the time.

So this chapter explains the law of agency.

2.1 Preliminary

We use the standard legal terms:

'agent' meaning the agent,
'principal' meaning the person represented by the agent,
'third party' meaning the other party to any contract, and
'authority' meaning the powers which the law gives to the agent.

Do not confuse the authority of directors and managers to run the organisation – internal management authority – with the external authority of an agent to commit that organisation when dealing with others. The two can overlap but are not the same. External authority is the power to act on behalf of the principal to create or change its legal rights and obligations in relation to third parties. Typically this power is exercised when making or managing contracts.

Using Commercial Contracts: a practical guide for engineers and project managers, First Edition. David Wright.
© 2016 John Wiley & Sons, Ltd. Published 2016 by John Wiley & Sons, Ltd.

2.2 Almost anyone can be an agent

The principal must have the ability/capacity to make the contract, but the agent does not need it, because he simply represents his principal. As a result, a child, for instance, can make a contract as an agent even though he cannot make it himself. As an example, a young girl can only make a *voidable* contract to buy shaving cream because it is not a necessary for her (see Chapter 4 below) but she can make a *binding* contract to so do when acting as an agent for her father.

2.3 Agency is created by consent

For agency to exist there needs to be consent between the principal and the agent. Normally this will be by some form of agreement, but in some cases the law will also be prepared to imply consent. So an agent can be made in the following ways:

by express agreement;
by implied agreement;
by operation of law; or
by 'estoppel'.

Then there are two other situations that may also occur – the ratification by a principal of the actions of an agent acting without authority, and the problem of the 'undisclosed principal'.

Agency created by express agreement between the parties

This is by far the most common situation.

In domestic life, it might be informal. If I ask my next door neighbour to buy me some sliced bread when she goes shopping and she agrees, then she will be my agent. In the commercial world the agreement can still be informal. The directors might simply ask one of their number to deal with a specific transaction or with a particular client, or a department manager might assign a particular area of responsibility to one of his people, and so on.

Agreement might be more formal. The board might appoint someone as a sales or procurement manager, giving him a formal variation to his contract of employment or a letter of appointment setting out areas of responsibility and levels of authority.

Then there might be a contract. One company might enter into an agreement with another to act as its representative or sales agent in a particular area or for a particular product, or might appoint an insurance broker or solicitor.

Where there is a contract, it is the substance that matters. The terminology or form does not. What is important is that it must give one party the authority to act as an agent on behalf of the other as principal in a way which will change the rights and obligations of the principal towards a third party. If it does not do this then no agency is created. But if it does create that authority, then the party given that authority will be an agent for the other.

Commercial agency agreements will usually be in writing, and there are some situations in which they must be in writing, or may need to be in the form of a deed or power of attorney. Any agent handling a transfer of ownership or declaration of a trust in any property must be appointed in writing (section 53 of the Law of Property Act 1925), and any agent who is to enter into a document that needs to be signed as a deed on behalf of his principal must also be appointed by deed, as must an agent who is to act for a trustee (the Powers of Attorney Act 1971 and the Trustee Act 1925).

And of course any agent being appointed to act as a personal representative in a situation where a person is unable to act for himself, usually because he is abroad for some period, or suffering from serious illness or incapacity, must be appointed by the Office of the Public Guardian. Previously this would have happened under the Enduring Powers of Attorney Act 1985; now it would be done by a Lasting Power of Attorney under the Mental Capacity Act 2005.

Agency created by implied agreement between the parties

An agent can be appointed by the implied agreement of the principal. This is uncommon but happens where the principal has not asked the agent to act for him, but accepts, or is presumed to accept, what the agent has done on his behalf.

A practical example might be the case where a landlord will be expected to pay for emergency repairs ordered by his tenant in the case of storm damage to the roof of a rented property (see also the next paragraph). Other classic examples are the power of a wife to 'pledge her husband's credit' when buying household necessaries and the statutory duty of the agent of a finance or insurance company to act as the agent of the client seeking finance or insurance cover as well as his own client.

Agency created by 'operation of law' or by 'necessity'

This is a very early piece of law. It was developed before the modern law of quasi-contract/restitution and implied authority, and is now largely unnecessary except in a very few circumstances.

Where emergency action is required in relation to goods or property, usually to protect them from loss or damage, and it is impossible to contact the owner for his instructions/authority, the law will allow any person in possession of the goods to protect/preserve them as an agent for the owner (so that the owner is responsible for the resulting cost). The only conditions are that the agent must –

- have made proper attempts to contact the owner if at all possible;
- act in the interests of the owner, and not from any other motive;
- only take action that is in the best interests of the owner; and
- do no more than is prudent.

Obviously in this modern mobile-phone world this does not happen often. But an example might be the need for a warehousing company to take emergency action to protect goods being held in storage in the event of a fire.

Agency of necessity will normally only be allowed where the agent is dealing with vulnerable or perishable goods or livestock. Two classic examples are –

Notara v. *Henderson* (1872): action taken/expense incurred by the captain of a ship to protect vulnerable goods being carried by his ship from loss, and

Great Northern Railway v. *Swaffield* (1874): the stabling of a horse at a nearby livery yard by a railway company when the owner failed to collect it at its destination.

(Similar statutory powers are given to the emergency services, such as the fire brigade and police, to take possession of and deal with private property in an emergency situation.)

In practice, agency of necessity can produce two situations – where the person in possession or control of the property needs to make a contract on behalf of the owner, or where he needs to take some other action, such as storing goods at his own premises.

In the first case, if he is not already an agent of the owner (for example, the GNR in *Swaffield*'s case) then clearly he can only have power to make a contract on behalf of the owner under the rules of agency of necessity. If, however, he is already an agent of the owner (the captain of the ship in *Notara*'s case) then he will have a general power to act in an emergency as a part of his normal powers as an agent or bailee, so that there is no real need to use the principles of agency of necessity at all.

In the second case (see *Prager* v. *Blatspiel etc*. (1924)), if he is already an agent of the owner the rules of implied authority (see below) will probably apply, as well as those of agency of necessity, so that he can charge for his work anyway.

But if the person acting is not an agent of the owner then it is settled that no agency of necessity can be created. The only exception to this is the law of marine salvage, where the principle applies that salvage of a ship or cargo entitles the salvager to payment by the owner.

There are two solutions to this problem. One is to rely on the law of bailment, which will usually entitle the person in possession to payment for his work or the costs of acting to preserve the owner's property or interests. The other is to include an appropriate clause in the contract to allow for emergency situations.

Agency created by 'estoppel' – or 'holding out'

'Estoppel' is simply lawyers' jargon. It comes from old English and means 'stopping *yourself*'. Sometimes the expression 'holding out' is used instead. Estoppel can both create an agent and create apparent or ostensible authority for an agent.

In terms of the creation of agency, the principle is very simple. If the principal, by actions or words, even if unintentionally, informs a third party that someone is authorised to act as his agent and that person then acts as his agent, the principal will be bound by what he has done. The principal will have stopped himself denying the statement that he made or denying that he intended that statement to be taken at face value.

Two examples –

> So far as third parties are concerned, whenever a company has a managing director he is automatically held out as having the authority to bind the company in any trading contract, whatever his actual authority might be – see *Hely-Hutchison* v. *Brayhead Ltd* (1968).
>
> A company entered into an agreement with a representative in Brazil to assist in developing the market and identifying potential customers for its products in Brazil. The agreement did not give the representative any authority to make any contract on behalf of the company. During a visit to Brazil by the sales director of the company, the representative introduced him to several potential customers. It later became clear that they had understood what the sales director had said during their meetings to mean that the representative did have authority to act on the company's behalf.

It is rare in normal commerce for agents to be created in this way. It is much more common for the authority of an agent to be increased by estoppel, and the principle will therefore be considered in more detail below under the rules of apparent authority.

2.4 Retrospective agency by ratification

The principle is an anomaly, in that it does not fit properly with the rest of the law on agency, but is well established.

In real life, ratification is most likely to happen where a negotiator has for some reason found it necessary to go beyond his instructions in order to reach agreement on his principal's behalf. (See also the section on implied authority.)

It is obvious that there is a considerable overlap between the different doctrines that can apply at the fringes of authority.

If a person who is not authorised to act as an agent agrees a contract with a third party for a principal, the principal may, when he becomes aware that the agent has made that contract on his behalf, accept what the agent has done by ratifying the contract.

For ratification to take place there must be:

- a person without adequate authority who has nevertheless acted as an agent on behalf of a named or ascertainable principal;
- a principal who has not given authority but is prepared to accept what has been done on his behalf;
- a third party who knows or has been informed by the agent that he is acting for that principal but does not have the proper authority;
- agreement between the agent and the third party on the terms of a contract
- that would be binding if made with authority; and finally
- an act of ratification by the principal.

The agent must act properly in the circumstances but not have authority. This may arise in two ways. First, he may have authority but agree terms that are outside that authority. This is not unusual in normal commercial negotiations. Alternatively the agent may have no authority, but find himself in a position to agree terms. In both cases he must believe that the principal would accept what he is able to agree.

The normal rules as to capacity apply. The principal must have the capacity to make the contract at the time that the terms are agreed, and the capacity of the agent is immaterial.

The principal must not have given authority, but must be prepared to accept what the agent has done.

If the principal is not prepared to accept what the agent has done then there will be no contract and the principal will have no liability to the third party. However, the agent has to be careful to advise the third party of the situation, or he risks committing the principal to a contract on the basis of apparent authority (see below). (Or the third party might even be able to claim damages from the agent on the basis that he has exceeded his authority.)

The third party must know that the agent is acting on behalf of a known or ascertainable principal, and must be prepared to enter into a contract with that principal.

Ratification by the principal must take place as soon as the principal has full knowledge of the circumstances. Ratification will be any statement or action that confirms that the principal does accept the contract made on his behalf.

2.5 The undisclosed principal

This is a very unusual corner of the law that has existed for two centuries. The law is reasonably clear, though the underlying principles have never been properly sorted out. It usually occurs in dealings with land, property, shares and so on.

Where an agent:

- acts within his actual authority (see below) and makes a contract
- as agent for a principal
- who has the capacity to make the contract
- without disclosing to the third party that he is acting as agent
- so that the third party believes that the agent is acting for himself
- then tells the third party that he has been acting as an agent for the principal
- the principal can take over the contract in place of the agent
- provided that the third party agrees.

If the third party does not agree, the agent remains personally responsible for the contract.

As an example of when it might apply, think of the situation where a buyer wishes to purchase shares in a company anonymously, possibly in a pre-take-over move, and asks a merchant bank to buy the shares on his behalf. The owner of the shares will be willing to sell the shares to the bank or to any other person. Once he knows the facts, the seller can choose to either stay with the merchant bank or transfer the contract to the principal. If the principal is a substantial organisation the seller might accept the transfer of the contract, but if the principal is less substantial the seller might prefer the merchant bank as a buyer, because the bank will clearly be able to pay the price.

2.6 The duties of the agent to the principal

The duties of the agent are very largely common sense. But the agent is in a position of trust as far as his principal is concerned, and therefore is someone who must act 'in good faith' to deserve that trust. So a high standard of conduct is expected.

The law does not really bother about the third party when dealing with the responsibilities of the principal and agent. Except where the relationship of the agent to third parties is regulated by consumer protection law to prevent selling abuses (the imposition of the seven-day 'cooling off' period for example) or in the case of a *del credere* agent (see below), the law sees the agent purely as an extension of the principal, and therefore as owing no obligation to the third party at all.

A *del credere* agent is only found in international trade and is now extremely rare, having been replaced by more modern forms of security for payment, such as the letter of credit or bank guarantee. The *del credere* agent would be a broker, acting on behalf of a seller. In addition to arranging the sale, the agent would, for an extra commission, give his personal guarantee to the seller that he would receive payment in due course.

Performance

In the commercial world the agent will normally have a contract with the principal, either an agency contract or a contract of employment. In that case the first duty of the agent is to comply with the terms of his contract, so long as it is neither illegal nor impossible nor would expose him to personal danger. He must also always act in the best interests of his principal.

An agent who has agreed to act without any contract or without payment does not have any obligation to perform, since there is no consideration for that promise, but if he does act then the other duties listed below will apply to him.

Personal performance

The principal has appointed the agent to act on his behalf. He has not selected someone else. Therefore the agent has an obligation to carry out his duties himself. He must not employ a subcontractor, or delegate the duties to someone else.

Of course there are qualifications.

The principal may agree, or even require, the agent to use another person to carry out all or part of his duties.

Then it may be a necessity or normal trade custom or business practice to do so. For instance, an agent for the sale of a house might employ an estate agent to advertise the property and a solicitor to carry out the conveyancing.

Finally, where the agent is a company or partnership, then the actual work would have to be carried out by an employee of the company or by a partner or employee of the partnership. In such a case, unless the agreement names a particular person who is to be responsible for the work, the agent will be entitled to select whichever person he wishes.

Skill

The agent must act with the proper level of skill. This will vary from person to person. An unpaid agent is expected to show the level of skill to be expected of the normal person. The agent under a contract is expected to show whatever level of skill is required by the contract or the normal level of skill to be expected of a person in his position or within his profession. An agent who is an employee is expected to act with all the skill that he has (the normal level of skill expected of any employee).

Obedience

The agent must obey the instructions of his principal, even if he disagrees with them or considers that they are not in the best interests of the principal. The only situations in which he may legally refuse to carry out any specific instructions of his principal are where it is impossible, dangerous or illegal to do so.

Accountability and conflict of interest

These are two quite separate and distinct duties, but may be considered together. They are the clearest examples of the good faith required from the agent. They can be very simply stated.

The agent must give full details to the principal of his actions on the principal's behalf, and must account for and transfer to the principal all monetary and other benefits that he has received while acting as agent, unless the principal has agreed/agrees that he may keep them.

Then he must not allow any conflict of interest between himself and his principal to arise, again unless the principal has already agreed to this, nor use his position to enrich himself at the expense of his principal (even unintentionally) without the principal's consent.

There are many examples of the application of this principle by the courts, ranging from cases where an agent employed to buy something for his principal has sold his own property to the principal, to cases where an agent has accepted commissions or bribes in respect of contracts that he is handling. Note that the honesty or dishonesty of the agent is not the issue, though of course a dishonest agent will be treated far more severely. Even an honest agent who has acted in such a way as to provide a considerable benefit to his principal must still account for any benefits that he may have himself received as a result and must avoid any potential conflict.

(Commercially the principal may accept that the agent will receive benefits or that a conflict of interest may appear to arise while the agent is acting for him. Often the person best qualified to represent the principal will be someone who already knows the market or problem well. That person will almost inevitably already have other clients, and the principal will be happy to accept that.)

Confidentiality

The agent must maintain confidentiality in respect of the affairs of the principal at all times, both during the period of his agency and afterwards.

2.7 The rights of the agent

Indemnity

The agent is entitled to be indemnified against all proper costs and expenses incurred while carrying out his duties, including any items which the agent may strictly not be liable to pay, such as tips/gratuities. He is also entitled to be indemnified against any losses and liabilities that he may have incurred.

If the agent is an employee, these costs will be dealt with by the principal in accordance with its normal procedures for payment of employees' expenses. Where the agent is working under a contract it is usual for the contract to set out how these costs are to be dealt with.

An unpaid agent will need to reclaim them from the principal.

Payment

Where the agent is to be paid, he is entitled to payment either on the basis stated in the contract, or if no basis has been fixed, then at the normal rates for that work, or if there are no normal rates that are applicable, then a reasonable amount. Of course it is usual in practice for the contract to deal with this in some detail.

2.8 The agent's authority

Preliminary

The real extent of the authority of the agent will largely depend upon the knowledge and understanding of the third party.

Employees of an organisation will behave responsibly and professionally almost all the time, and exercise authority with care. Nevertheless, the organisation will always need to manage the use of agency authority. It is a four-part process. The organisation must:

- set up clear rules for the exercise of authority;
- ensure that anyone who has any authority is aware of its limits;
- ensure that third parties know who is authorised to act on behalf of the organisation (and possibly also what the limits to their authority are); and
- keep breaches of the rules to the absolute minimum (because any breach of the rules tends to destroy their validity).

Authority must always be considered in two ways.

The principal and agent will always be concerned with the authority agreed between them, what is described below as actual authority. In any dispute between them the law also will only be concerned with that actual authority.

But if there is a dispute between the third party and the principal or agent, then the authority of the agent is not what the principal and agent know it to be, but what the third party reasonably considers it to be.

If the third party has been told by the agent or the principal, or perhaps even by another appropriate party, what the actual authority of the agent is, then both principal/agent and third party will have the same understanding. But if the third party has not been informed of the actual authority of the agent, then he may decide for himself what the agent's authority is, probably based upon the principles of implied or apparent authority. This is the level of authority that the law will then enforce.

The reasoning behind this is very simple. The agent's implied or apparent authority can exceed his actual authority. If an agent has exceeded his actual authority but not his implied or apparent authority, then he may have agreed something that is acceptable to the third party, but not acceptable to his principal. If the principal can refuse to comply, the third party will suffer. If the principal has to comply then the principal will suffer. The loss will not be the fault of either the principal or the third party – but it is the principal who has appointed, and can control, the agent who has actually caused the loss. Therefore the principal should bear the loss. See further *Lloyd* v. *Grace, Smith & Co* (1912) below.

So, implied or apparent authority may and probably will take precedence over actual authority.

Therefore we can look at the authority of the agent on the basis of:

1. Actual authority.
   This is the authority actually agreed between the agent and principal – it will be 'specific', 'normal' and/or 'incidental'.
2. Implied authority.
   This is the authority that comes from the position of the agent/employee within the principal's organisation.
3. Apparent, or ostensible, authority.
   This is the authority that the agent has as a result of a specific task that he has been given to perform.

Actual authority

The actual authority of the agent is the authority given to him by the principal. It is created by the agreement of the parties. Sometimes this agreement may be by a contract between them, for example, a contract to employ a commercial agent, estate agent or solicitor. Sometimes it may be an appointment, as when a company gives one of its employees the specific authority to enter into or manage a particular contract for the organisation.

The problems of actual authority –

Lloyd v. *Grace, Smith & Co* (1912) L was an elderly lady who employed GS, a firm of solicitors, to manage her affairs. One of her contacts with GS was through A, a senior clerk, who visited her whenever papers needed to be signed, and so on. During one particular visit when there were several items to be signed, A tricked her into signing a deed transferring some houses that she owned to him. A then sold the houses and disappeared with the (considerable) proceeds. L sued GS. The court held that A was employed by GS to visit clients and get papers signed. That was what he had done. So even though he was committing a fraud, he had acted within his actual authority, and GS were responsible for what he had done.

There are three aspects to authority that need to be borne in mind. The concepts are actually simple, but the terminology used by the law can sometimes become rather confusing and confused.

i) First there is specific authority, the authority given by the principal to the agent. This may be particular, say to negotiate or manage one or more named contracts, or general, say to negotiate or manage any contracts that might arise. It may be stated in an agreement or implied from the appointment of the agent to a specific position within the principal's organisation.

ii) Second there is customary authority, sometimes called 'normal' or 'usual' authority. This is what the professional agent would be expected to do. A stockbroker or insurance broker would do deals for his client within the market; an estate agent would advertise property, and so on.

In a number of nineteenth-century cases, agents selling animals or farm produce on behalf of their principals were held to be entitled to give customers guarantees of the quality of the items that they were selling which were then contractually binding on the principals, because this was simply normal practice.

In other words, the principal always needs to understand clearly the normal practice of any professional agent that he intends to use.

An example –

Morris v. *C W Martin & Sons Ltd* (1966) M sent an expensive white mink stole to B, a furrier, for cleaning. B sent it on to CWM who was expert in dealing with mink. The fur was stolen by an employee of CWM. It was held that the employee was employed to handle furs, so that he was acting within his actual authority. Also B was working within his normal authority, by selecting an expert to do the actual work. So CWM were liable for the loss; see Chapter 14 below for more detail.

iii) Third there is incidental authority. This is what the agent might do as a normal part of his work. An estate agent would take photographs, put up signs, and so on.

For instance, in the case of *Panama Developments (Guildford) Ltd* v. *Fidelis Furnishing Fabrics Ltd* (1971) a company secretary hired cars, nominally on behalf of the company but actually for his own use. It was held that car hire was one of the normal administrative tasks that he might be expected to do from time to time. Therefore, even though he had misused his authority (and had in fact acted fraudulently), the company had to pay for the hire of the vehicles.

Implied authority

Implied authority comes from position within the organisation. Every organisation relies on it to get things done. It tells the outside world whom they can deal with. It gives people job titles, such as contract supervisor, purchasing manager or sales director. The job title announces to third parties the position which the person holds within the organisation, and with that position the implied authority that it is reasonable to expect that person to have.

So the actual authority of a sales director, for instance, might extend to all sales contracts up to a certain size/price, with contracts above that needing approval by the managing director or board of directors. But the implied authority of a sales director would normally extend to all sales contracts of whatever size.

Consider this situation:

A company appointed someone to a new position as 'European Sales Manager' for oil/petrochemical industry contracts within a number of Western European countries. His actual authority was set at £2 million. Contracts above that value had to be approved by the board of directors before signature. He was involved in negotiations for a contract worth slightly below £2 million. At almost the last minute the customer asked for a change to the specification which increased the price to over £2 million. The sales manager requested a delay of 48 hours for contract signature so that he could obtain board authorisation to sign the contract at the new price. The third party accepted the delay. However, he told the sales manager that it was customary for his suppliers' sales managers to have authority of around £5 million.

What would have been the position if the sales manager had decided not to risk asking for 48 hours delay? There would then have been two courses open to him. He could have signed the contract immediately, but at the same time informed the customer that his acceptance of the contract would need to be ratified by his company; see above. Alternatively he could simply have signed the contract without revealing that he did not have actual authority to do so. If he had done this, would the contract have been effective – given the statement by the customer that it was normal practice for people at his level to have the authority to sign contracts of that size? He would clearly have exceeded his actual authority, but almost certainly would not have exceeded his implied authority. Therefore the contract would have been effective. But the sales manager would have had a problem with his directors, unless they were completely confident in his judgement.

And of course there was another question which the board needed to consider. Should the level of the manager's authority be increased?

Apparent or ostensible authority

As we have seen, implied authority is the reasonable third party's view of the authority of directors/managers within an organisation when doing their normal work. Apparent authority covers everyone else.

The theory is straightforward. If the organisation puts one of its own employees into a position, or allows one of its employees to put himself into a position, in which it is reasonable for a third party to assume that he has the authority to act as an agent for the organisation, and the employee does then act, that action will bind the organisation.

> In *Freeman & Lockyer* v. *Buckhurst Park Properties (Mangal) Ltd* (1964) the directors of a company knowingly allowed one of their number to manage the business for a considerable period, in effect permitting him to act as managing director, although he did not have any authority to act independently. He made several contracts on behalf of the company. The company was sued for payment for work carried out to his instructions and was held liable. Although the board had given him no formal authority, by allowing him to continue to act on the company's behalf, the board had led the supplier to believe that he was authorised to act as agent for the company. The result was that he had apparent authority even though the contracts that he had entered into were extremely ill advised.

Apparent authority in the hands of an experienced employee is generally not a problem. The difficulty lies in the fact that it can apply to almost any employee of the organisation who comes into direct contact with a third party either during contract negotiation or during the performance of a contract. Of course there are strict limits to the apparent authority of the vast majority of employees, because it must always be reasonable for the third party to assume that the employee has a specific type and level of authority. Nevertheless, this is an area which will always merit serious management attention.

Two practical examples –

> A purchasing manager is due at a meeting to discuss a contract with a supplier. Unfortunately he is detained upon more important business and asks one of his buyers to go to the meeting in his place. The manager has actual authority to sign any contract up to £250,000. But he does not wish to give nearly so much authority

to the buyer in this case. He therefore tells the buyer to give the contract to the supplier if he can agree a price of £20,000 or less, but that if he cannot get that price then he must refer back to the manager for a decision.

The result is that the buyer now has actual authority of £20,000. However, the supplier will not know this. As far as he is concerned the buyer is there to discuss and agree the terms of the contract with him. Therefore, unless the buyer, or someone else on behalf of the organisation, has already told the supplier, or tells him during the discussion, of the limits on the buyer's actual authority, the supplier will assume that the buyer has the authority to agree whatever he does actually agree. This is the buyer's apparent authority.

In the negotiation the buyer achieves agreement on a price of £21,000, but achieves useful concessions from the supplier in terms of equipment specification and terms of payment that make the overall supply package commercially very acceptable. What should the buyer now do? To comply strictly with his instructions he must now inform the supplier that he has no authority to complete the contract but must refer what has been agreed back to his manager for ratification (or possible further negotiation). That is what a junior or inexperienced buyer must do. However, that might risk the supplier also reserving his right to renegotiate the deal, and so a more experienced buyer might use his own judgement and accept the contract, knowing that he was exceeding his actual authority but that what he was doing would be automatically accepted by his manager.

The organisation may create apparent authority almost by accident. It can be very much wider than anything intended by the organisation. This is because the law gives any third party the right to assume, when it is reasonable for him to do so, that the organisation is in proper control of those who act or appear to act on its behalf, and therefore the right to rely on commitments that they may give in the name of the organisation. It is therefore perfectly possible that a person may be able to commit the organisation to a contract while having no proper authority to do so from within the organisation.

Purchase orders used within an oil refinery clearly stated that only the purchasing officer had the authority to make a variation to the work to be carried out under the order. But an oil refinery is a very large place and a contractor working on the site could be sometimes literally miles away. Also, a refinery is not an easy working environment. There are serious risks involved, and a 'good' contractor who could be relied on to carry out high-quality work safely was a very valuable resource and therefore not to be squandered.

The problem was that supervisors within the refinery who needed a minor repair carried out urgently would, for the best of reasons, 'borrow' a contractor's people working on the site to get the repair done. The contractor would carry out the work and the purchasing officer might know nothing about it until the contractor submitted his invoice for the extra work.

The purchasing officer is now in a dilemma. If he were to follow the wording of the purchase order, he should refuse to authorise payment to the contractor. That would result in a very unhappy but valuable contractor, and as soon as word got round (as it would inevitably do), several other unhappy contractors as well. In addition he would have an unhappy supervisor, and probably an unhappy production manager/director. If he were to authorise payment everyone would be happy, but he would have lost control of his contract budget and would have confirmed to the contractor that the statement in the purchase order that only the purchasing officer had the authority to give a valid order variation was meaningless, and that any supervisor had the authority to instruct him to carry out extra work under any order. Purchasing officers felt that they had little choice but to authorise payment.

Of course everyone was treating the problem as one of a variation to an existing purchase order. Was this correct? Maybe it was. Certainly both parties were acting on the convenient basis that this was an item of work to be carried out as an addition to the scope of work under an existing contract. But on the facts the supervisor might actually have made a quite separate verbal contract between the refinery and the contractor – so that the contractor should have invoiced for the work separately from the existing purchase order. (In other words, the supervisor would then have had very different apparent authority.)

Perhaps there was not even a contract for the extra work, but merely an obligation for the refinery to pay the contractor a reasonable sum for carrying out the work under the law of restitution (see below), because the work had been done in response to a request from someone who had no authority to place an order for the work, but who made the request knowing that the work would cost money to carry out and would clearly be of financial benefit to his organisation.

(You can see just how much profitable fun the legal profession could have had with that situation if a purchasing officer had refused payment and the matter had ended up in court.)

An agent with no authority at all or the agent that never was

This is the result of the peculiar case of *Watteau* v. *Fenwick* (1893). A owned a public house, but got into financial difficulty and sold it to F. F retained him as manager and licensee, so that his name remained over the door. F gave A the authority to buy only beers and mineral water for the public house. A had no authority to buy anything else. A then bought a consignment of cigars from W, something which he had done in the past. W was not aware that A had sold the business to F and believed that A was still the owner. A could not afford to pay – and F refused to do so. W sued F. Clearly A had at best exceeded his authority or at worst had no authority at all. But the problem was that W made it quite clear that he did not know that A was acting as an agent for the owner – he believed that A was the owner. He was not aware that F existed. So the doctrine of the undisclosed principal could not apply. The doctrine of apparent authority could not apply either. The contract could not be void for mistake either, because W knew the person who was buying his goods.

Nevertheless the court decided that F was liable to pay W because A was acting as his agent. (Nowadays of course W would claim payment under the law of restitution.)

But the decision was clearly highly suspect at the very least. It has been 'justified' on the basis that it is an example of a creative judge creating a new theory and the court's willingness to extend the law by creating unusual authority to obtain a just result in a situation where the facts did not quite fit the existing theory. It appears in all the textbooks, but has never been followed in any other case.

2.9 The commercial agent

The Commercial Agents (Council Directive) Regulations 1993 (SI 1993/3053, as amended by SI 1993/3173/483 and SI 1999/201) is the UK legislation that incorporates the requirements of EU Directive 1986/653 into UK law (published in the Official Journal (1986) L382/17).

A commercial agent is described as an independent (self-employed) person or organisation who or which is authorised to negotiate, or to assist in negotiating, contracts for the sale/purchase of goods for his principal, either in respect of a particular product or more usually within a particular territory or area. The essential nature of the agent's function is that of business development, finding appropriate buyers or sellers, establishing relationships, dealing with enquiries and so on. He is not simply a salesman or buyer.

Many companies depend on agents of this kind when selling outside their own home country or countries.

The regulation simply provides that the commercial agent will have all the normal duties of an agent towards his principal, but that at the termination of any agreement with his principal, the agent will be entitled to receive a payment of commission in respect of contracts made by the principal after the termination of the agreement but which have come about as a result of the work done by the agent.

The legislation sets out to reward commercial agents for future contracts when their agreements are terminated for any reason by giving them the right to compensation for the loss of future commission that they would have earned had the agreement remained in force.

As to what sort of person or organisation will be a commercial agent, two cases illustrate the issues –

> *Parks* v. *Esso Petroleum Co Ltd* (2000) involved the termination of an agreement for the running of a self-service garage/service station, with the fuel being sold by

P as an agent for Esso. The court decided that the function of a commercial agent involves being involved with the 'negotiation' of contracts. This implies skill, not just selling. P was not a commercial agent.

P J Pipe & Valve Co v. *Audco India Ltd* (2005) An agent in India conducted preliminary discussions, got his principal on to bid lists, and helped to manage the relationship with buyers during the contract negotiations. The agent did not agree prices or terms of contract and was not authorised to sign contracts. He was held to be a commercial agent.

While there can never be a definitive rule as to the amount/period of compensation that might be reasonable, the UK courts have adopted a policy of taking the commission earned by the agent during the period of two years before termination as a starting point when considering what might be reasonable. This is no more than a rule of thumb however. In the case of *Lonsdale* v. *Howard & Hallam Ltd* (2007) the court held that a much lower level of compensation should apply when the business of the principal was in serious decline at the time of termination.

3

The Organisation

There are four basic types of commercial organisation.

These are –

- the private person (as a sole trader);
- the partnership;
- the company;
- the corporation and other organisations.

3.1 The private person

The simplest contracting entity is the private person. He buys for personal use or consumption for himself or his family. However reckless, stupid or impulsive he may be, he has complete freedom to manage his own lifestyle and income.

He will deal with other private individuals on a basis of equality, but whenever he deals with a commercial organisation he is protected as a matter of public policy, mostly by legislation, from the consequences of unfair practices. A person is protected –

- when a consumer, by consumer protection legislation;
- when a child, or especially vulnerable for any other reason, by the law on capacity (see below); and
- when an employee, by employment protection legislation.

Using Commercial Contracts: a practical guide for engineers and project managers, First Edition. David Wright.
© 2016 John Wiley & Sons, Ltd. Published 2016 by John Wiley & Sons, Ltd.

3.2 The sole trader

But when the individual goes into business everything changes. He no longer has the protection of employee/consumer law. He is expected to behave just like any other business. He is now a sole trader.

The sole trader (or individual proprietor, or self-employed person) is the simplest form of commercial organisation. In the UK, as in most other countries, this type of organisation will normally be very small, from a one-man business up to an organisation employing a handful of people. Larger businesses are possible, but rare. (In Germany by contrast sole trader businesses may be much larger.) In the eyes of the law, the sole trader is his business. He owns all the assets of the business and is personally responsible for its debts. He has the sole right to manage, to trade, and to take the profit from the business, and must also bear the losses. He is responsible for any contracts made by the business, whether made by him or made on behalf of the business by any employee or other person acting as his agent. He is also expected to have normal commercial judgement and skills.

> *L'Estrange* v. *Graucob Ltd* (1934) Mrs L'Estrange owned and ran a small café/snack-bar. She signed a contract to buy an automatic vending machine from the defendant, without reading the small-print terms of the contract. There was no pressure by the salesman, or any misstatement by him as to what the contract terms were. The machine failed to work properly and she claimed damages. The company was protected by a clause in the contract. The court dismissed her claim – she was in business and had signed the contract so she was bound by it. Whether she had read it – or would have been competent to have understood it if she had - was immaterial. As she was in business she was presumed to have understood the need to act with normal commercial forethought. She had signed the document and must be deemed to have understood that it contained contract terms. If she was prepared to accept those terms without reading them first, then she must be prepared to accept the risk that they might be to her disadvantage.

> Contrast this with *Curtis* v. *Chemical Cleaning and Dyeing Co* (1951). In this case C took an expensive dress to be cleaned. The shop asked her to sign a form that contained a clause excluding all liability for any damage done. C quite properly asked about the form and was told that it merely limited liability if 'any sequins fell off during cleaning'. C accepted and signed the form. The dress was ruined. C claimed and the court rejected the exemption clause on the basis of the misrepresentation of its meaning by the shop.

3.3 The partnership

A partnership is defined as 'a relationship between persons carrying on a business in common with a view of profit' (section 1 of the Partnership Act 1890). The normal term used to describe a partnership is a 'firm' (section 4 of the Act).

In the UK the vast majority of partnerships are in the professions. But any group of people, companies or corporations, because they are also 'legal' persons, combining to carry on a business or business venture may be a partnership. Many consortia and joint ventures which carry out major contracts or projects on a collaborative basis are actually partnerships within the meaning of the Act.

Members of the partnership may be salaried or may share in the profits of the venture, and the partnership may have employees and own property, just as any other business.

The Partnership Act does not require the partners to enter into any formal relationship, and some partnerships at the most basic level may exist without any formal written agreement between its members at all.

However, for very practical reasons most partnerships will be based upon a written partnership agreement, setting out how profit or loss is to be shared, how new partners may be brought into the partnership or existing partners may leave, and so on, and finally, how the affairs of the partnership are to be managed.

The 'English law' partnership has no separate legal identity. A contract with it is a contract with all the people who are members of the partnership at the time the contract is made. Each of those members will be personally fully liable on a joint, but not several, basis, for any failure to perform or any breach of the contract terms. (In other words, any legal claim must be made against all the partners collectively, not against a single individual partner, and each partner will only be liable for his share of any damages.) The liability of the partners is unlimited in extent.

For a 'Scottish law' partnership the position is different. The partnership does have separate legal personality. Also, the partners have both joint and several liability for breach of contract. (In other words, any claim can be made against any partner or partners, and each partner can be liable in full for any damages.)

Under both English and Scottish law the partners are jointly and severally liable for 'wrongs' – for damages for tort under English law, or delict under Scottish law.

If there is no written partnership agreement, or the agreement is silent, each of the partners will have equal authority within the partnership and also full actual authority to act as agent for the partnership. However, especially within larger partnerships, it is usual for there to be a partnership agreement that provides for the management of the partnership by the more senior partners and for contracts to be made only by one or more of them. In that case partners outside the managing group may not have any significant authority, but the third party must be made aware that this is the case. Of course, they will still have a considerable degree of power delegated to them to deal with particular contracts being carried out by the partnership.

The Act sets out the position as follows in section 5 –

Every partner is an agent of the firm and his other partners for the purpose of the business of the partnership: and the acts of every partner who does any act for carrying on in the usual way business of the kind carried on by the firm … [shall] bind the firm and his partners unless the partner … has in fact no authority to act for the firm in the particular matter, and the person with whom he is dealing knows that he has no authority or does not know or think him to be a partner.

Then the Act states, in section 8, that any member of the partnership has the power to act as an agent for the partnership when acting within his 'usual or apparent authority', and that if it is agreed that there is any restriction on the authority of a partner, then in the event of any act by him which is beyond the limits of his authority, the restriction will (only) be binding with respect to persons having notice of the agreement. As far as others are concerned, the restriction will be of no effect. In other words, the normal principles of implied and apparent authority apply.

In addition, the partnership may also delegate power to its employees or others in the same way as any other organisation.

A variant of the partnership is the limited partnership, governed by the Limited Partnership Act 1907. Under the Act a partnership may include one or more 'limited partners' who have invested in the partnership and who share in the profits of the business. However, they take no part in the work of the partnership and have virtually no management powers, and certainly no agent's authority. In contrast to the unlimited liability of the full partners, the liability of the limited partners will be limited to the extent of their investment in the event of a claim against the partnership. The authority of the full partners and employees will be as stated above.

Finally, there is the Limited Liability Partnership, created by the Limited Liability Partnership Act 2000. This is an incorporated body with separate legal personality from those of its partners, and a registered office, not dissimilar to the German GmbH. Under the Act, its authority and management structure is the same as that of any other partnership.

3.4 The company

When dealing with any company that you do not know well, remember that the company can change almost everything. It can have and change a trading name. It can change its own name. It can change its registered office, its trading address(es) and its owners/shareholders. It may go into liquidation and be replaced by another company. The only thing that it cannot change is its registered number, the number by which it is known to Companies House.

The most common commercial trading organisation is the limited company, which can be a plc, a public limited company, or simply a limited company that may be private or part of a group of companies. The company has a legal personality that is separate from that or those of its owner(s). It also has a much more closely defined internal management structure than a partnership does.

For virtually all UK companies the two key documents are the memorandum of association and the articles of association. The memorandum describes the actual and potential businesses in which the company may engage. For comments on the memorandum see Chapter 4. The articles set out the basis upon which the company will be controlled and managed. In general terms the company will be owned by its shareholders, but managed by the board of directors on their behalf. The articles of association of UK trading companies will generally be broadly in accordance with the appropriate standard set of model articles provided under the Companies Acts. The current versions are set out in the Companies (Tables A to F) Regulations 1985 SI 1985/805. These are A for companies limited by shares, B for private companies limited by shares, C and D for companies limited by guarantee, E for unlimited companies, and F for plcs.

Of course, the tables are updated whenever a new Companies Act is passed, and most UK companies have been in existence for many years, and will as a result have Articles based upon earlier versions. However, successive updates have left the relevant sections essentially unchanged. The current versions deal with management authority in articles 70 to 72. (In the previous version the equivalent sections were articles 77 and 78.)

1. Subject to the provisions of the (1985 Companies) Act, the memorandum and the articles and to any directions given by special resolution, the business of the company shall be managed by the directors who may exercise all the powers of the company. [...] The powers given by this regulation shall not be limited by any special power given to the directors by the articles and a meeting of directors at which a quorum is present may exercise all powers exercisable by the directors.
2. The directors may, by power of attorney or otherwise, appoint any person to be the agent of the company for such purposes and upon such conditions as they determine, including authority for the agent to delegate all or any of his powers.
3. The directors may delegate any of their powers to any committee consisting of one or more directors. They may also delegate to any managing director or any director holding any other executive office such of their powers as they consider desirable to be exercised by him. Any such delegation may be made subject to any conditions the directors may impose, and either collaterally with or to the exclusion of their own powers and may be revoked and altered.

What these articles do is to give the ultimate internal management authority within the company to the directors, as a committee. The directors also have the actual authority to deal with third parties on behalf of the company, and to authorise others to so deal as well. In addition the directors can also appoint executive directors, departmental managers, etc., giving them the authority to run various areas of the company's business, and delegate power to them to deal with third parties as necessary for that purpose. In turn those executive directors and managers will have the authority to delegate power to employees within their departments, and so on, subject to whatever limits they consider appropriate.

The problem for the company of course is that while these limits to the actual authority/power of people within the company are known to the individuals concerned, they may not be known to third parties unless those third parties have been told of them or have come to be aware of them in some way, through a continuing business relationship for example. (See Chapter 2.)

3.5 The corporation and other organisations

Finally, there is a wide variety of other organisations. There are 'co-operatives', and mutual societies, such as the smaller building societies. (Major building societies are now generally companies.) Building societies, for instance, are governed by the Building Societies Acts, which require the society to have a constitution and rules, the rules fulfilling the same function as the articles of association within the company, and with much the same result.

Then there is the 'corporation'. This category includes a vast range of different organisations, from the BBC, through health service bodies and other trusts, learned and professional institutions, authorities, agencies, universities, and local government councils, to national government departments. They are far too diverse to try to deal with individually. They can be either sole or aggregate. The best example of a corporation sole is that of a government department, headed by its minister, a secretary of state. An example of a corporation aggregate might be a local government council, headed by a 'mayor and corporation'. Most corporations were originally created by or draw authority from an Act of parliament, from which comes a document setting out how the corporation is to be run, or from a charter, a constitution, statutes or rules of one form or another, which create an authority/management structure in the same way that the articles create an authority/management structure for the company.

The corporation, like the company, has separate legal personality, and authority to deal with third parties devolves from the head(s) of the corporation down through the organisation in the same way as for the company. The corporation also remains in being even though the head of the organisation may change. A contract to carry out work for a government department is strictly speaking a contract to carry out work for the minister in charge of the department, whoever he or she might be at any particular moment. For our purposes it is enough to say that the principles of authority within the corporation are broadly the same as for any other trading organisation.

3.6 Finally

The commercial organisation is expected to be able to look after itself. Put simply it is required to equip itself with the skills that are essential to carry out its business properly. It may choose to employ those skills or to buy them in from external advisers, but it fails to do so at its peril.

The organisation is expected to be commercially skilled. When the organisation makes a contract, the law expects the organisation to have considered whether or not that contract represents a reasonable commercial bargain. If it later turns out that the contract is not a reasonable bargain, or that through some chance what was a reasonable bargain at the time has turned out badly, then in the eyes of the law that is simply a risk that the organisation has to accept. The law will not allow the organisation to avoid the results of a contract that has become a commercial disaster, if for no other reason because that would be unfair to the other party to the contract.

The organisation is expected to be commercial. Every organisation is expected to be able to act in a proper commercial manner. Even the most altruistic of charities, for example, is expected to be able to approach its contracts in a proper manner, and to build the normal commercial precautions into those contracts. If the organisation fails to do so it can expect no help from the law.

The organisation is expected to be technically skilled. Whatever the field in which the organisation operates it is expected to have the technical skills necessary to carry out any contract that it has accepted. The only exception to this rule is where circumstances intervene to make it impossible to carry out the contract.

The organisation is expected to be professional. The professional organisation will employ people who are suitably trained for the work that they have to do and, where necessary, people who have the required qualifications for any specific functions which they may be required to carry out. It will also have the necessary procedures and systems in place to ensure that its employees have appropriate guidance as to how to carry out their functions properly.

The organisation is expected to know the law. If an organisation fails to carry out a contract properly because it has made a mistake and misunderstood the law or legal requirements in the UK (or any other country), and therefore failed to comply with those requirements, then that is the responsibility of the organisation. This is of course self-evident. But some areas of law are always developing, often developing at quite a high speed, the current hot spots being 'health and safety', 'pollution control' and 'environmental protection', and keeping up to date on changes to law, possibly in a number of different countries, can sometimes be a problem.

The organisation is expected to be contractually skilled. As we will see below, the law puts considerable emphasis upon the 'correct' interpretation of the words actually used by the parties both when making their contract, and when carrying it out. So the organisation needs to be competent in writing its contracts, and also competent in reading those contracts after they have been made. The principle is that when a professional organisation makes a contract, that contract will use the words that the organisation intends to use. The organisation will have used those words because it knows what those words mean, and it is happy to live with the benefits/obligations package set out in those words. It must be prepared to carry out the contract in accordance with those words.

The organisation has bargaining power. The theory is that all commercial organisations have power to negotiate the terms of their contracts. Therefore, when the organisation accepts a contract, that contract represents a deal which has been negotiated by the organisation and agreed to as acceptable. Therefore the organisation must live with the terms that it has agreed. In the modern world of 'standard-form' contracts, and where a major supermarket chain or utility or IT company may have almost infinitely more bargaining power than a small company, this principle may look rather weak. However, it is true in one sense – that in commercial situations the organisation, however small, does have the power to refuse to accept the contract. This of course leads us to …

The organisation has freedom of choice. The organisation will always have the power to refuse to take the contract. So if it does take the contract then it has consented to its terms, and the idea of consent is fundamental to the theory of contract.

4

Making the Contract Part 1 – The Requirements

This chapter and the next look at the rules that apply to making a contract. In themselves the rules are simple, but they can cause problems in real life. This is because it is all too easy to make mistakes when trying to apply them.

The requirements are:

- parties with the capacity to make a contract;
- intention to create a legal relationship;
- certainty of the terms and capability of performance;
- legality of the contract;
- consideration; and
- dealings that satisfy an analysis in terms of 'offer and acceptance' (which is examined in the next chapter).

4.1 Parties with 'capacity'

When the contract is made, the parties to the contract must have 'capacity', that is the ability to make that contract. *Prima facie*, everyone is presumed to have the capacity to enter into a contract, unless it can be shown that he lacks capacity for some particular reason. There are three aspects to this rule that are of importance.

The person must exist

The person must exist at the time that the contract is actually made. This looks self-evident. Clearly a non-existent person cannot make a contract.

Using Commercial Contracts: a practical guide for engineers and project managers, First Edition. David Wright.
© 2016 John Wiley & Sons, Ltd. Published 2016 by John Wiley & Sons, Ltd.

But there are a number of cases where someone has jumped the gun and placed contracts for a company before it has been fully incorporated. This usually happens when there has been an unexpected delay in the incorporation/registration of the company. Where this happens, logically the company cannot be bound by the contract. Neither can the company ratify or adopt the contract as an undisclosed principal (see the principles set out in Chapter 2 above). So who should pay the supplier?

Of course the usual solution is that as soon as the mistake comes to light the 'contract' is immediately replaced by a new and correct contract.

But what if one of the parties seeks to avoid liability? Under the European Communities Act 1972 and the Companies Acts 1985 (sections 36(4) and 36C) and 1989, which implemented Article 7 of the EU First Directive on Company Law, the position is that if any contract is made by any person purporting to act in any capacity (whether as agent, promoter, director or employee) on behalf of a company that is not yet incorporated or able to trade, then he will be personally liable for performance of the contract, unless there is an express statement in the contract to the contrary – see *Phonogram Ltd* v. *Lane* (1982).

Furthermore, if the contract expressly provides for its enforcement by the company or gives a benefit to the company, then the company may be able to enforce it in due course under the terms of sections 1(1) and 1(3) of the Contract (Rights of Third Parties) Act 1999; see Chapter 13 below.

Young persons (or 'minors')

Certain issues apply to the case of children and young people below the age of majority – now set at 18 years. Very small children are incapable of entering into a contract. Once a child is old enough to understand what a contract involves, he can make that contract. For instance, a child who is old enough to understand the concept of paying to buy sweets can buy them. And so on. Then young people up to the age of majority are able to make contracts. But the law divides their contracts into two kinds, contracts for 'necessaries', and other contracts.

Contracts for necessaries are binding, both on the young person and on the other side, provided that the terms of the contract are reasonable. (Even a contract for necessaries will be voidable by the young person if the terms are oppressive or too harsh.) 'Necessaries' have always been given a restricted meaning. First, they include the needs for life, food, clothing, accommodation, transport, professional services or medical care, and so on, both for the young person and for his or her dependents. Everything in this category will count as necessaries, provided they are appropriate for the young person's lifestyle. Luxuries are never (except perhaps in the case of a teenage multimillionaire) necessaries in the eyes of the law. One of the classic cases on the subject concerned whether a dozen fancy waistcoats were necessaries for a Cambridge undergraduate – they were not: *Nash* v. *Inman* (1908).

Then, necessaries also include contracts for education, employment and apprenticeships. However, see Chapter 12 on 'restraint of trade' below.

Other contracts, for non-necessaries as they are termed, are binding on the other party, but may be cancelled by the young person. Where the contract is for the purchase of something permanent, such as shares in a company or an interest in land, then the contract is binding upon the young person, but he has the option to cancel that contract at any time until he reaches his majority, when the option will cease. Other contracts are not binding on the young person unless he ratifies them after reaching majority.

But of course the vast majority of such contracts will not be cancelled. Every child or teenager spends money on luxuries.

Finally, note also that by definition a young person cannot enter into a binding trading contract in his own name, because it is a contract for non-necessaries. (Of course, the same young person can own all or the majority of the shares in a company, and act as an agent for that company to place contracts on the company's behalf.)

People who are mentally disordered

In contracts for necessaries the position of someone suffering from mental disorder is no different to that of a young person. The contract is binding provided that it does not contain harsh or oppressive terms, when it is invalid. Other contracts will also be binding unless the disordered person did not know what he was doing, and the other side realised that this was so, in which case the contract may be cancelled. The Mental Health Acts 1959/1983 and the Mental Capacity Act 2005 have now put into place a reasonably comprehensive system for ensuring that the affairs of seriously mentally disordered patients are properly administered on their behalf.

People under the influence of alcohol or drugs

Contracts made by people who are completely intoxicated are void because the intoxicated person does not know what he is doing and therefore is incapable of giving consent. This is also probably also the case where the person is not completely intoxicated, but is sufficiently intoxicated not to have a proper understanding of the contract that he is entering into. However, lesser intoxication, sufficient to affect judgement but not enough to affect understanding, will not invalidate the contract.

Others

Finally, there are also statutory restrictions on the ability of persons who are bankrupt and not yet discharged, and of persons who have been convicted of various offences, to enter into contracts for certain types of non-necessaries.

The capacity of the organisation

The problems caused by organisations seeking to avoid liability under contracts which they had made but that were not properly within their capacity to make (what were called actions *ultra vires*, actions that were 'beyond the powers' of the organisation), exercised the law for many years, and a considerable number of cases dealt in erudite fashion with the issues involved. The problem could arise in two ways. The organisation might have entered into a transaction of a type that was not permitted to it, or the organisation might have entered into a transaction that was allowed, but in a way that was in breach of the organisation's rules.

Actions of a type not permitted

Where the organisation concerned, such as a university or building society, is governed by a charter or by rules, any act that falls outside the charter or rules will be beyond the power of the organisation. However, to protect the position of the third party, the contract will be valid but the organisation may be liable to regulatory action.

For a company subject to a memorandum of association the position used to be different. The memorandum set out the purposes for which the company was formed and the areas of activity in which it could engage. Any activity outside those areas, and they were often very limited indeed, was beyond the powers of the company. It was therefore void, which of course meant that any loss would fall on the third party, who was usually completely blameless. (Even today some companies will still have restrictive memoranda.) The reasoning for this was that as the documents concerned were documents of public record, the third party must be taken to be aware of them.

The situation has now been very largely remedied by the European Union. Article 9 of the First Directive on Company Law in 1968 required member states to bring this doctrine to an end. This was done in the UK by, successively, the European Communities Act 1972 and the Companies Acts 1985 and 1989. For normal trading companies the point is now dealt with in section 35 of the 1985 Act as amended by section 108 of the 1989 Act.

In relation to any other party who deals in good faith, all companies and statutory corporations are automatically deemed to have the power to make any contract that they have actually made, together with any other (non-contractual) acts that they may have carried out (such as the proper authorisation of a security) and they are also deemed to have properly authorised anyone who has acted on their behalf in doing so.

For charitable companies the position is dealt with in section 111 of the 1989 Act, and now sections 65–68 of the Charities Act 1993, and is slightly different. Here the *ultra vires* doctrine may still apply to non-trading contracts, but no longer applies to normal trading contracts (that is contracts for adequate consideration in terms of payment for the supply of goods or services) or to any contract made when the third party did not know that the company with which he was dealing had failed to follow the proper procedures or was a charitable company.

Incorrect procedure

Where the transaction was 'merely' made in breach of the proper procedure the transaction was outside the actual authority of the agent who made the transaction on behalf of the organisation, but might be within the powers of the organisation. The principle was therefore established in *Royal British Bank* v. *Turquand* (1856) that where the third party was ignorant of the breach of proper procedure he was entitled to assume that the organisation had properly followed its internal procedures, and that therefore the contract was valid.

However, where the other party was aware that internal procedures might not have been properly followed then he would not be able to enforce the contract unless before entering into the contract he had queried the situation in a proper manner and received assurances that all was correct from the person with actual authority to manage the business of the organisation.

(The policy behind both these rules was that the investments of shareholders in the organisation should be protected at all costs, even if this meant sacrificing the interests of those who traded with that organisation, perhaps because in the Victorian era the encouragement of investment was of prime importance.)

The result was very unsatisfactory because it left the trading partners of the organisation to carry a risk which was to a large extent outside their control.

British Bank of the Middle East v. *Sun Life Assurance Company of Canada Ltd* (1983) The Bank was involved in dealings with Sun Life which required the issue of a type of security by Sun Life that was perfectly normal for a bank but unusual for an insurance company, and would therefore normally only be authorised at senior level within the company. The security was issued by a branch manager. The Bank was concerned by this and wrote to the General Manager asking for confirmation that the security had been properly authorised. Assurance was given but by Sun Life's regional office, not the General Manager. Not satisfied with this the Bank again wrote to the General Manager and again received assurance but again at regional office level, not from the General Manager. The Bank reluctantly accepted this assurance. Later Sun Life refused payment against the security on the basis that it had not been properly authorised. The Bank sued and lost, on the basis that as the Bank knew that a security of this kind would need to be authorised by the General Manager it should not have accepted the assurance at regional management level but should have continued to demand an assurance from the General Manager.

This situation has also been remedied by the sections in the Companies Acts 1985/1989 referred to above.

4.2 Intention to create legal relations

'Intention' is not so much a requirement for the contract, as a presumption of law. This is that –

- the parties to a 'non-commercial' agreement, that is any agreement between private persons, do not intend their agreement to be legally binding unless it is completely clear that the agreement is meant to be legally enforceable, but
- the parties to a commercial agreement will always intend that agreement to be a binding contract unless they have clearly stated in the agreement that the agreement is not to be legally enforceable.

Non-commercial agreements

Many agreements between private individuals will be legally binding contracts, even sometimes agreements between members of the same family. A parent might employ a son or daughter in the family business, for example. A husband and wife might agree a marriage contract, the settlement terms that should apply in the event of their later divorce. Anyone advertising a second-hand car or camera in the columns of the local paper or on the internet will expect any sale which follows to be legally binding and so will the buyer. But there are many agreements made in the course of normal family life which are never meant to be legally binding. Parents will support children going into further education or university, or a husband and wife will agree on a monthly allowance for domestic expenditure, or how the household expenses are to be shared between them. These arrangements may become contracts but only if it is absolutely clear that this is what the parties intend.

This is best illustrated by some of the classic cases.

Husband and wife –

In *Balfour* v. *Balfour* (1919) B was a civil servant working in Sri Lanka. At first his wife accompanied him. She then suffered ill health and they agreed that she should remain in England and that he would pay her a monthly allowance. After some years he stopped paying, because he was in fact now living with someone else. It was decided by the Court of Appeal that the agreement was not a legally binding contract and that therefore the wife was not entitled to insist on continuing payment of the allowance. (In other words, she should sue for maintenance.)

In *Merritt* v. *Merritt* (1970) on the other hand some 50 years later husband and wife separated. The husband left home to live with another woman. He agreed to pay his wife a monthly allowance on the basis that she should accept responsibility for paying the mortgage on their house which was in joint names. He also agreed that when the mortgage ended he would transfer his share in the ownership of the house to her. Mrs M insisted that this agreement was put in writing. This was done. The actual words used were 'in consideration of the fact that you (Mrs M) will pay all charges in connection with the house … I will agree to transfer the property in (it) to your sole ownership'. When the mortgage was paid off he refused to transfer ownership. She successfully sued him for his refusal. The court held that the fact that the agreement had been put into writing and the circumstances in which it was done showed clear intention to create a contract.

Parent and child –

In *Jones* v. *Padavatton* (1969) J and P were mother and daughter. They agreed that P would leave her job in America and come to London to study for the bar (which would normally take about three years) and that J would support her while she was doing so. Initially J paid P an allowance, but two years later J bought a house, part of which was rented to tenants, the rest being occupied by P who lived on the rents from the property. Three years after that J and P quarrelled and J claimed possession of the house. It was decided by the court that the agreement was not intended to be legally binding and J was given possession.

It was, however, clear from the judgments of the court that this was a very marginal decision. (However, if this arrangement had been held to be a contract, it could only have been for a reasonable duration, and since P had already been drawing an income from her mother for over five years, and had still only passed a small part of the Bar examinations it was perhaps unlikely that the outcome would have been any different.)

Non-domestic agreements –
In *Simpkins* v. *Pays* (1955) S lodged with P and her granddaughter. The three submitted a joint entry each week to a newspaper competition. The entry was submitted in the name of P, but all three parties shared in the costs of the entry. One entry was successful and P won £750. P refused to pay anything to S, and S sued P for a third of the winnings. It was held that this was intended to be a mutual arrangement between the parties for a joint venture to which all three contributed in the expectation that any prize that was won would be shared between the parties. The arrangement was therefore a contract, and P was ordered to pay S his share.

Commercial agreements

The principle is very simple. In the commercial world every agreement will automatically be a contract, unless it is completely clear from the words used that the parties did not intend their agreement to be legally binding. Of course the number of cases that have come before the courts in which one side or the other has tried to avoid liability by arguing that it was not the 'intention' of the parties to make a binding contract is legion. However, the number of times that anyone has been successful in avoiding liability on this basis is very small indeed. Two cases will serve to illustrate the point.

> *Rose and Frank Co* v. *Crompton (JR) & Bros* (1925). This case arose out of an agreement made in 1913 between RF, a New York importer, and C, a UK manufacturer. The agreement gave RF the right to sell C's products within the US. It contained the following clause 'This arrangement is not entered into nor is this memorandum written, as a formal or legal agreement, and shall not be subject to legal jurisdiction in the law courts either of the United States or England, but it is only a definite expression and record of the purpose and intention of the parties concerned, to which they each honourably pledge themselves'. After the agreement had been in operation for several years, C terminated it without giving proper notice, and also refused to supply goods against a number of outstanding purchase orders from RF. RF sued for breach. The court held that the agreement clearly stated that it was not to be capable of enforcement by legal action, and so C was not liable for termination. But the purchase orders made under it were all valid contracts. C was therefore liable for breach of contract in failing to supply goods against them.

> Contrast this with the case of *Edwards* v. *Skyways Ltd* (1964). E was a pilot employed by S. S decided to reduce the number of its employees and gave E notice of termination of his employment. E was a member of S's contributory pension scheme. After negotiation S offered E two options, to take a paid-up pension or to accept the repayment of his contributions to the scheme, plus what S described as an 'ex gratia' payment roughly equal to S's own contributions. E accepted this second option. S repaid his contributions but then withheld the ex gratia payment. When challenged S claimed that as they had agreed only to make the payment on an ex gratia basis there was no contractual obligation to pay it. E sued for the payment and his claim was upheld. The court held that the use of the words 'ex gratia' was not enough to defeat the presumption that in a commercial agreement there was an intention that the agreement should create a legal relationship. Only the clearest statement of intention would be enough.
> (Also see the case of *Carlill* v. *The Carbolic Smokeball Company* in Chapter 5.)

4.3 Certainty of terms

This can be a problem in the commercial contract. The principle is easy to understand. There can be no contract until the parties have agreed all the terms necessary for a contract. The practical problem is that commercial deals often go through several stages as the parties work their way towards agreement. This means that some matters may not be decided until long after the rest of the deal has been made.

> *May & Butcher Ltd* v. *R* (1934) This case arose out of an arrangement between MB and the government Disposals Board for the purchase by MB of government war surplus stock (tents). The arrangement was set out in a letter from the Board to MB, which arrangement had been accepted by MB.
>   The letter said –
>
> 1. The (Board) agrees to sell and (MB) agrees to buy the total stock of old tentage …
> 2. The price or prices to be paid, and the date or dates on which payment is to be made by (MB) to the (Board) for such old tentage shall be agreed upon from time to time between the (Board) and (MB) as the quantities of the said old tentage become available for disposal, and are offered to (MB) by the (Board) …
>
>   The parties fell out and the Board then refused to sell to MB, claiming that the terms set out in the letter did not constitute a binding contract. The court agreed. For there to be a contract the parties must have agreed everything that is necessary for a contract. Nothing must still be left to be agreed between the parties.

In other words, an agreement to agree is not legally enforceable, except perhaps as an agreement to make a reasonable attempt to try to reach final agreement.

(If, of course, the letter from the Board had said that MB should pay a reasonable price [in accordance with section 8 of the Sale of Goods Act; see below Chapter 9] or that the price should be fixed by an independent third party, then it would have been a contract.)

The terms that the parties have agreed must also be clear.

The practical difficulty here is that the commercial contract will often have to deal with topics or areas where precision is very difficult to achieve. Many contracts also need to resort to the use of such terms as 'to a good commercial (or engineering) standard', 'in accordance with best practice', 'due diligence', and so on, terms that are used to define ideas that are well understood by the professional but intrinsically difficult to define in precise legal detail. Agreeing complex terms takes time and effort. As a result the final details of the deal are often postponed.

An example of the difficulties that can be caused by this is the following –

> *Scammell* v. *Ouston* (1941) O was in discussion with S for the purchase of a lorry. The basic terms of the purchase were that part of the price would be paid by O

trading in O's old vehicle, and that the rest of the price would be covered by hire purchase. However, no precise terms for the hire purchase element were agreed. Later S refused to supply the new vehicle and O sued.

The first difficulty was that there was no proper purchase contract. At the end of the discussions O merely wrote to S ordering the vehicle and stating 'this order is given on the understanding that the balance of the purchase price can be had on hire purchase terms over a period of two years', which was accepted by S. The court disposed of this by recognising that very often the commercial businessman does not write contractual documents with the precision that would be expected of a lawyer.

But the second difficulty was that there were a number of different variations of hire purchase terms that could apply. Some were more onerous than others. There was no guidance in the 'order' or the correspondence or in any previous dealings between the parties as to which terms would apply. The court therefore held that the terms were not certain and that as a result there was no contract.

The way that the law on certainty of terms has developed is, to say the least, both confused and confusing.

The difficulty is that most of the cases arise from situations where a deal had been struck that has operated perfectly satisfactorily for some time, but then has broken down. One of the parties will want to terminate the arrangement, usually because there is a better deal to be had elsewhere, or because it has become too expensive to continue. The other will want to keep the arrangement going, or to recover compensation for the loss of a deal that has become much more profitable to him than it was to begin with.

It will then become clear that the words used in the contract are less than perfect, with the result that the court has to struggle to make reasonable legal sense out of imperfect commercial words. The consequence has been that judges have tended to give a series of reasonably commonsense decisions, but have justified those decisions in ways which conflict with each other. So the decisions are reasonably consistent, but the law as explained in those decisions is not.

But the law is that, wherever possible, even if the wording of the agreement is less than perfect, it will still be a contract so long as the parties have agreed upon at least the minimum terms necessary to create a contract. If necessary the terms written into the agreement will also be interpreted in the light of the normal commercial or technical standards, customs or understandings within the trade or business to which the agreement relates. In general terms the law will treat any reasonable commercial agreement as a legally binding contract if it is possible to do so.

This approach is shown in two classic cases, *Hillas* v. *Arcos* and *Nicolene* v. *Simmonds*, which demonstrate what can be achieved in a situation where the parties have given the court rather more to work with than in *Scammell*'s case.

Hillas & Co Ltd v. *Arcos Ltd* (1933) H was a wood merchant. A was a Russian state trading company that sold lumber and wood products from the Soviet Union. The parties made an agreement for the supply by A of softwood to H. The agreement was that in 1930 Arcos would supply to H '22,000 standards of softwood goods of fair specification' at A's list price less 5%. H was also given options for the purchase of 100,000 standards on the same terms during each of the years 1931 and 1932. The contract was performed properly during 1930, but in 1931 the price of softwood increased sharply. As a result A was able to sell its entire UK allocation of timber elsewhere at a good price. So, when H sought to exercise its option for that year, A refused to supply, claiming that the language used to define the timber to be supplied was imprecise, so that there was no binding contract. The court rejected this argument. First, the parties had not had any difficulty in carrying out the contract in 1930. Second, it was clear from the expert evidence produced by H that any professional within the trade would know what was meant by 'a standard of fair specification'. (Essentially it meant an average tree trunk. Of course some tree trunks would be larger than average and some would be smaller but that would balance out.) Therefore, the 1931/1932 options were binding.

Nicolene Ltd v. *Simmonds* (1953) N wrote to S offering to buy 3000 tonnes of steel reinforcing bar. S replied in writing accepting the offer, but then adding the words 'I assume that we are in agreement that the usual conditions of acceptance apply'. There were in fact no 'usual conditions of acceptance' common in the trade, or previously used by the two parties, so that no expert evidence could be produced as to what the words might mean. S then failed to supply and N claimed damages for his failure. S defended the claim on the basis that the wording of the agreement was uncertain. The court rejected this claim. There was a world of difference between a clause that had an uncertain meaning, and therefore still needed to be clarified before there could be a contract and words that were merely an addition to an otherwise complete contract and had no meaning at all. Because the words had no meaning they were to be disregarded. The contract was therefore completely binding and S was held liable for damages. (Of course, the court might perhaps also have taken into consideration that it was S, the defendant, who inserted the meaningless words into the contract in the first place.)

4.4 Capability of performance

As well as being certain the terms must be able to be performed. This seems self-evident. An impossible contract can never be performed and so can never provide consideration. But in the case of *Harbutts 'Plasticine' Ltd* v. *Wayne Tank and Pump Co Ltd* (1970) a contractor was held to be responsible for failure to carry out work – installing plastic piping that was specified as having the heat/fire resistant characteristics of steel – in accordance with a specification that was actually impossible at that time. The court simply held that as a professional contractor that was his problem.

4.5 Legality

A contract that is 'tainted with illegality' may produce a whole variety of different results, depending upon whether it is deliberately or accidentally illegal, whether one party or both or neither knows that it is illegal, the nature of the illegality, and whether it is for an illegal purpose or is for a legal purpose but is carried out in an illegal way or by someone who is not properly authorised or licensed to carry it out.

The subject is beloved by academic lawyers, is very complicated and is considered in detail in Chapter 12 below. At this point it is simply enough to say that a contract for a purpose that is illegal will either be wholly illegal, and therefore of no effect, or will be capable of being rendered ineffective by one or other of the parties once the problem is discovered. A contract that is mostly in accordance with law but which contains an illegal element (such as an employment contract containing an illegal restraint of trade clause) will be liable to having the illegal element struck out. A contract for a legal purpose but which is performed in an illegal way will be valid.

4.6 Consideration

The law of contract is not concerned with enforcing a promise. (If a promise to do something is to be enforceable on its own, that is, without being included in a contract, then it needs to be put into the form of a deed, or to be brought within the provisions of the law of restitution; see Chapter 15 below.) The law is concerned with enforcing a bargain.

Consideration is the jargon term used to mean the bargain element required for a contract. The legal rules are straightforward. For there to be a contract, each of the parties has to do something or promise to do, or not to do, something for the other party.

But the route that the current law has taken to get here is confused. The reason for this is the difference between English common law and continental civil law systems. Under civil law systems a simple promise made 'deliberately and for good reason' is enforceable. This theory has its attractions for judges who want to make decisions that are just and fair whenever they can.

As a result there have been various attempts to change the rules on consideration to make deserving promises enforceable if at all possible. However, these attempts to change the principles of consideration are now in the past, well in the past, and do not need to be considered any further.

Subject to the exceptions listed below, consideration is any act or forbearance or promise to do something or to refrain from doing something that is either to the actual or apparent benefit of the person to whom consideration is being given or to the actual or apparent detriment of the person giving the con-sideration. It must have some value, but that value may be minimal or out of all proportion to the value of the consideration provided by the other side. It must be performed at or after the time that the contract is made. Acts or promises made before contract can only be valid consideration in special circumstances (see below).

(An example of a promise to refrain might be along the lines of 'If you will agree not to sue X for breach of a particular contract I will agree to carry out repairs to the equipment … ')

The time of performance

Consideration may be, in legal jargon, either executed or executory. Executed consideration is consideration that takes place at the time, or virtually at the same time, that the contract is made. Executory consideration is a promise in the contract to act or refrain from acting after the contract has been made. They are best explained by examples.

Typically, executed consideration will happen in a consumer contract. Say a buyer purchases a book in a shop, and pays for the book in cash. The contract is made at the time of purchase. The purchase may consist of a small number of stages, probably the buyer naming the book that he wants, an assistant produc-ing a copy and naming the price, the buyer agreeing to buy and paying the price, and finally the assistant handing the book to the buyer. Essentially the contract is made at the time when the buyer agrees to buy the book. However, at virtually the same time the buyer will pay the price. Also at virtually the same time the assistant will deliver the book to the buyer. Payment of the price is the buyer's consideration for the contract. Delivery of the book to the buyer is the shop's consideration. Both of these acts, payment and delivery, take place at the same time as the contract is made. This is executed consideration by both parties.

In a commercial contract on the other hand, perhaps a purchase order for equipment, the purchaser will order the goods from the supplier, who will accept the order. The supplier will agree to deliver the goods within an appropriate period and the purchaser will agree to pay for those goods when they are delivered. In this case the purchase order will become a contract when it is accepted, but nothing else happens at that time. Instead, the contract contains two promises – a promise by the supplier to deliver the goods to the purchaser, and a promise by the purchaser to make payment for those goods when they have been delivered. Neither of these two acts of supply and payment will take place until sometime after the making of the contract. All we have at the time of contract is two promises. These promises are executory consideration, one for the other.

Finally, a contract may contain both executed and executory consideration. Using the example of the purchase of the book, if the assistant had told the buyer that the book was not in stock, but that he could obtain a copy for him within a few days provided that the buyer paid the price in advance, and this was accepted by the buyer, then the contract would contain both types of consideration. The buyer would have given executed consideration because his consideration was given at the time that the contract was made, whereas the bookshop would have given executory consideration, a promised to obtain and supply the book at a time after the contract had been made.

Past consideration

Acts or promises made before a contract can only be valid consideration in special circumstances. 'Hard cases make bad law', and the rules on 'past consideration', as it is called, arise from a series of just such hard cases. In one typical example a widow lived in a house that now belonged to her children as the result of her late husband's will. She spent considerable sums on the property. Years later, her children agreed 'in consideration of your carrying out certain alterations and improvements to the property, we hereby agree …' that she should be paid a substantial sum. They later refused payment and the court had no choice but to reject a claim by the widow because there was no consideration for the promise to pay (*Re McArdle* (1951)).

Another example was *Roscorla* v. *Thomas* (1842) in which a guarantee given before the sale of a horse that it was 'well-behaved' was held to be unenforceable.

For commercial contracts at least, the matter was largely remedied in a case in 1892.

Re Casey's Patents, Stewart v. *Casey*. S carried out work for C, the owners of a number of patents, helping them to manage the process of preparing documents for submission when applying for the patents and so on. C later wrote to S as follows, 'We now have pleasure in stating that in consideration of your services as the practical manager in working both our patents as above … we hereby agree to give you one third share of the patents above-mentioned, the same to take effect from this date …' Later C reneged on their promise and claimed that S had no right to a share of the patents because there was no consideration for their promise, since the work that S had done was completed before their letter was written, and was therefore 'past consideration', in other words, not consideration. The court held that commercial people do not work without expecting payment. It was clear that both sides always understood and intended that S should receive reasonable recompense for his work in due course. Therefore, that earlier understanding could provide consideration for the letter granting S a share in the patents.

In a later case (1980), the Privy Council applied this decision and Lord Scarman stated the rule as follows –

An act done before the giving of a promise to make a payment or to confer some other benefit can sometimes be consideration for the promise. The act must have been done at the promisor's request: the parties must have understood that the act was to be remunerated either by a payment or the conferment of some other benefit: and payment, or the conferment of a benefit, must have been legally enforceable had it been promised in advance.

Also see Chapter 15 below – the same recognition that commercial organisations expect payment for work also underlies the law on letters of intent.

Value of consideration

Consideration must have some value, or as the law puts it, must be 'sufficient'. First, it must be real, that is, it must have some value in the eyes of the law. That is why a promise made 'in return for your love and affection' or 'in return for not boring me with your complaints' is not a contract. For the same reason a promise to do something which both parties know to be impossible, such as to sell something which does not exist or cannot be manufactured, has no value.

Consideration does not need, however, to have any significant value. The law is not concerned with whether or not either side in the contract gets a good deal, or even an adequate deal, as long as the parties actually do make a deal. One of the classic cases on the point, *Bainbridge* v. *Firmston* (although the case was actually one of bailment), goes back as far as 1838. In that case the defendant made an agreement with the plaintiff to allow him to dismantle and weigh some machinery, in return for a promise, which he then failed to keep, to reassemble the machinery in good order. The court held that there was adequate consideration on both sides to create a contract. (Of course the statement of legal principle cannot be contested but one wonders whether the actual decision was entirely correct. The problem that the court faced was that the owner of the equipment could only get some recompense for the damage that he had suffered at the hands of the defendant if there was a contract between them.) In another equally classic case, the wrappings from bars of chocolate were held to be valuable consideration, even though they were then simply thrown away *(Chappell & Co Ltd* v. *Nestle & Co Ltd* (1960).

Also, everyone will be aware of the practice of using nominal consideration when writing documents such as confidentiality agreements, 'In consideration of the payment of £10 the receipt of which is hereby acknowledged …' The device of nominal consideration was in use at least as early as the 1840s.

Thomas v. *Thomas* (1842) The executor of a husband agreed with his widow that she should be allowed to remain in the matrimonial home, because her husband had expressed the wish that this should be so, and in consideration of the payment of £1 per year. The agreement was contested, but both counsel and the court accepted that the payment, though quite clearly completely inadequate as economic rent for the house, nevertheless was sufficient consideration for the agreement to constitute a contract.

Consideration must come, or 'move' in legal terms, from the person who promises, that is the person who makes the contract. In other words, an agreement between A and B that A will pay B £500 if C decorates A's office cannot be a contract between A and B. The reason for this is that it is not B, but a third party, who is giving the consideration to A, and there is therefore no contract. Equally it cannot be an agreement between A and C. If then C does decorate the office but A refuses to pay B, B will have no right to sue for breach of contract. (Of course there are ways round the problem. One is to make the agreement under seal or as a deed, when A's promise will be enforceable by reason of the document used. Another might be for B to promise that he will subcontract the work to C. That would be completely satisfactory consideration. A third solution would be for B and C to contract with A on a 'joint and several' basis.) See also the comments on the doctrine of privity of contract and the rights of third parties, Chapter 13 below.

For reasons of public policy, certain types of act or promise to act are 'insufficient' consideration. Essentially these relate to situations where the promisor already has a duty of some kind, so that he has to carry out the act anyway –

Where the promisor already has a public duty imposed by law. In the case of *Collins* v. *Godefroy* (1831) G promised to pay C, a policeman, for his inconvenience in attending court to give evidence. C had already been served with a subpoena requiring him to do so. The court rejected the claim for payment of the money.

On the other hand a policeman who gives information that leads to the arrest and conviction of a criminal is entitled to claim any reward. Also, the police are entitled to claim payment where they have agreed to provide any specific level of policing or protection for any particular premises or event, such as a football match. The leading case here is *Glasbrook Brothers* v. *Glamorgan CC* (1925), which related to an agreement that the police should provide a specific level of protection to a coalmine during a strike.

Where there is already a contract for the consideration between the parties. The classic case on this topic is *Stilk* v. *Myrick* (1809). A sailor had contracted to sail on a ship travelling to the Baltic and back to the UK. Halfway through the voyage two of the crew deserted and the captain promised extra wages to the other crew members if they would help work the ship back to the UK shorthanded. This meant anything up to a 50% increase in each man's watch-keeping duties. The court rejected the seaman's claim for the extra money on the basis that he already had a contract for the work of sailing the ship.

The reasoning in the case was impeccable, but the actual decision was shockingly bad. However, the case took place during the Napoleonic wars when the maintenance of strict discipline in both merchant and naval ships was a prime national consideration. In a case some 50 years later where the circumstances were very similar, except that the voyage was to the West Indies, a court had no difficulty in finding that where a ship had to be sailed seriously shorthanded, the increase in workload required of the crew was so considerable that their original contract obligations were discharged and they were therefore entitled to payment for that extra workload (*Hartley* v. *Ponsonby* (1857)). See also the comments upon the law of economic duress, Chapter 11 below.

A much more recent example of this is *Williams* v. *Roffey Bros & Nicholls (Contractors)* (1991). A subcontractor had got into financial difficulties in carrying out work and the main contractor agreed to pay the extra costs of completing work to avoid the collapse of the subcontract and the subsequent impact on his main contract. It was held that the benefit that he gained by the extra payment was valid consideration for its being enforceable.

Changes to payment

There is a difficulty with consideration where the parties agree to modify an existing obligation, in particular the payment of an existing debt. This is in a sense simply one application of the rule stated above. If A has an obligation to pay B £100, then a second contract stating that A will pay B only £50 in settlement of that debt cannot be valid because there is no consideration for the promise. Once a debt of £100 exists it cannot be changed. (This had been decided as early as 1602 in *Pinnel* v. *Cole*.) But there are two types of debtor. One is the reluctant payer who just does not want to settle his debt. The other is the debtor who is at least trying to pay, but is short of money.

> *Foakes* v. *Beer* (1884) Mrs B had obtained a judgment against Dr F for a sum of over £2000 (a considerable sum at the time). She then agreed not to enforce the judgment, but to give him time to pay. It took him some six years. Mrs B then claimed judgment interest on the money. Dr F objected but the court awarded the interest to Mrs B. She had a legal right to the money, and had agreed to allow Dr F several years to pay (and had not had Dr F declared bankrupt).

Another 'fobbing off the creditor' case was the following –

> *D and C Builders* v. *Rees* (1966) DCB were owed £482 by R for work that they had done. After several months, knowing that they were in difficulties, R offered them £300, saying that if this was refused R would not pay them anything. DCB reluctantly accepted the money but then sued for the balance and won.

But, especially in commercial life, someone may get into financial difficulty and will then need to achieve some compromise settlement of its debts, usually on the basis that a creditor or creditors will accept payment of less than the full amount of the debts owing.

> *Hirachand Punamchand* v. *Temple* (1911) Son T owed money to HP under a promissory note due to mature in some months' time. Father T offered immediate payment of a smaller sum 'in full settlement' of the debt and HP accepted the payment. HP then claimed the balance from son T. The court rejected the claim. Immediate payment by the father in cash was different to payment by the son under a note, which was not due for payment until later, even though it was of a smaller amount.

So, where there is an agreement between a debtor and a single creditor, it is now accepted that it is not possible simply to change a debt of £100 into a debt of £50 by a second contract. But it is permitted to change a debt of £100 into a debt of £50 by a separate contract, provided that the second contract also changes the way in which that debt is to be paid or the date on which that debt is to be paid. In other words, a debt of £100 payable on the first of August, for example, cannot be changed into a debt of £50 payable on the first of August, but it can quite properly be changed to a debt of £50 payable on the first of July or the first of September.

Where the agreement is made between a debtor and his creditors generally, what law calls a 'composition with creditors', the agreement will provide that each of the creditors will accept a proportion of the debt owing to him in full settlement of that debt. In effect each creditor will agree to refrain from suing the debtor for the amount actually owing to him in return for a guaranteed payment of a proportion of his debt. Therefore, each creditor will provide consideration to each of the other creditors, and also to the debtor. There is of course still one problem. It is difficult to see precisely what consideration is given by the debtor to the creditors. This is a subject that the courts have very carefully avoided discussing whenever the subject has arisen. Perhaps consideration could be found in the fact that when promising to pay all the creditors a set proportion of their debts, the debtor is in fact promising each creditor that he will not pay any other creditor a sum that is proportionately greater than the amount which he will actually pay to him. (The practical problems that could be created if a composition with creditors was not upheld are so serious that the court would have to find some way of upholding any composition that was challenged.)

5

Making the Contract Part 2 – Offer and Acceptance

The offer/acceptance process simply means that for there to be a contract, one side has to offer the terms and the other has to accept those terms in full and while the offer is still valid. It provides a series of simple and logical rules for working out:

- whether a contract has been made;
- when it was made; and
- what the terms of that contract are.

It works well when applied to simple buying and selling. It also works fairly well for normal commercial contracts.

But there may be many things that go into making a commercial contract, with various terms being agreed at every step on the way. There may be –

- previous trading contracts;
- overarching or 'frame' contracts;
- exploratory discussions and expressions of interest;
- formal or informal pre-qualification;
- feasibility studies;
- preliminary negotiations;
- budget price submissions;
- enquiry/invitation to treat;
- a 'budget offer';
- a tender, or offer, and perhaps a number of alternative tenders;
- post-tender negotiations;
- tender resubmissions or modifications;
- tender or price re-validation;
- acceptance.

Using Commercial Contracts: a practical guide for engineers and project managers, First Edition. David Wright.
© 2016 John Wiley & Sons, Ltd. Published 2016 by John Wiley & Sons, Ltd.

The offer/acceptance process does not take any account of most of these contacts. It concentrates solely upon the process itself, to the exclusion of everything else. So it is usual, and advisable, to insist on having a formal contract document, which separates out the terms finally agreed from everything that has gone before.

5.1 Making the deal

There are two questions that always have to be answered. Were the parties agreed on the terms of a contract? And, if they were agreed, did they then make the contract?

> *Tinn* v. *Hoffman* (1873) Two companies, one in the UK and one in the US, wrote to each other at virtually the same time, one offering to sell and the other to buy, the same goods at the same price. Then one company changed its mind. The other sued for breach of contract. The decision was that although both sides were ready and willing to do a deal they never actually managed to make one.

5.2 The objective approach

Law, especially commercial law, is not about whether anyone intends to make a contract. Instead it takes an objective view. Has a party acted in such a way that a reasonable person would believe that he intended to make a contract? What would others read into the actions or words that have been used? This can produce some surprising results.

One graphic example is the interesting and legally correct case of '*Bardell* v. *Pickwick*', as reported in *The Pickwick Papers*. Another is the following case –

> *GHSP Inc* v. *AB Electronics* (2010) GHSP, based in Michigan, was a supplier of electronic equipment for trucks to the Ford Motor Company in the US. It bought sensor units for truck accelerator control systems from AB. Due to a mistake by AB's Taiwanese subcontractor, a batch of faulty units was supplied which caused major problems of uncontrollable deceleration and 'engine stumbling'. This resulted in a product recall by Ford at a cost of some US$2 million, and claims by Ford against GHSP and by GHSP against AB. In fact, GHSP and AB had been in discussion about the terms of a purchase order for some months. GHSP's purchase conditions asked for unlimited liability in respect of any faulty equipment supplied. AB's sale conditions excluded liability. During the discussions AB had said that it would accept GHSP's conditions but only subject to a cap on liability, but AB had never proposed an amount for the cap and GHSP did not feel itself under any pressure to do so, so that the parties were effectively deadlocked on the point. Before delivery commenced GHSP sent a purchase order subject to its purchase conditions to AB and the manager of AB's accounts

department sent GHSP an 'acknowledgement of order', more to confirm that AB would deliver against the order than anything else. The court decided that the acknowledgement of order created a contract between the two, even though AB certainly, and GHSP probably, realised that the major issue of liability was still unresolved.

The court then decided that, as the units were delivered by AB within the UK, GHSP's conditions of purchase, making the contract subject to Michigan law, could not apply, and neither could AB's own conditions of sale. So the contract had to be governed by the basic law under the UK Sale of Goods Act, including the section 14-implied conditions of satisfactory quality and fitness for purpose (see Chapter 9 below) and of course without any limit of liability. The result was that AB was held liable for the full amount of the claim.

5.3 The stages

We have used the standard legal terms 'offeror' to mean the person who makes or possibly may make an offer, and 'offeree' meaning the person to whom that offer is or may be made.

There are four stages to consider: preliminary discussions, 'invitations to treat', the offer and the post-offer possibilities.

5.4 Preliminary discussions

The law puts no value on the various discussions and exchanges between the parties during the preliminary stages. They create no legal obligations between the parties, even though they may well be of substantial commercial or technical importance.

5.5 Invitation to treat

'Treat' is simply jargon, meaning 'discuss' or 'negotiate'. An invitation to treat is any act or communication which requests or invites an offer. (In the commercial world it will usually be in the form of an enquiry, often setting out in considerable detail the basis on which the offeree wishes to contract.) It is the usual starting point of the offer/acceptance process. It may be aimed at one particular offeror, at a selected group of offerors, or at the world at large.

The invitation can lay down a range of legally binding rules

First, it can lay down rules that apply to offerors. For example, it can specify the procedure, dates and times for submitting offers, the format in which offers should be submitted, accompanying documentation required, and so on. If it does so, any offers that fail to comply may be rejected.

Attempts to specify the method of submission are not usually successful. So an offer delivered by hand in good time in response to an enquiry stating 'delivery by first-class post' on a certain date was properly submitted and had to be considered (*Blackpool and Fylde Aero Club Ltd* v. *Blackpool B C* (1990)). But in the case of *Manchester Diocesan Council for Education* v. *Commercial and General Investments Ltd* (1970) where the enquiry specified only one very specific method of submission for offers, it was held that only offers submitted by that method had to be considered.

Second, it can lay down rules that apply to the offeree. If an invitation states that the lowest (or highest) offer properly submitted will be accepted, then the offeree is contractually bound to accept that offer. See *William Lacey (Hounslow) Ltd* v. *Davis* (1957).

An invitation to treat is not an offer

An invitation to treat is merely an indication that the offeree issuing the invitation is prepared to enter into a contract, but it is not itself an offer and is not therefore capable of being accepted. This distinction can be important.

> *Gibson* v. *Manchester City Council* (1979) In 1970 the Conservative City Council wrote to council house tenants with details of a scheme for the sale of council houses to their current tenants on favourable terms if tenants wished to apply. G completed and returned the printed reply form enclosed with the Council's letter. The city treasurer responded as follows: 'The Corporation may be prepared to sell the house to you at the purchase price of £xxxxx … ' He then set out details of the mortgage scheme available, and continued: 'This letter should not be regarded as a firm offer of a mortgage. If you would like to make a formal application to buy your council house please complete the enclosed application form and return it to me as soon as possible'. G filled in and returned the form but asked for the price to be reduced due to repairs needed to the property. The Council replied that these repairs had already been allowed for in the price that it had quoted and G then replied: 'In view of your remarks I would be obliged if you will carry on with the purchase as per my application already in your possession'. At this point Labour took control of the Council from the Conservatives and stopped the scheme, except where the Council was already legally bound to sell. The Council therefore ended negotiations with some 350 tenants including G. This was therefore a test case. The court decided that neither of the letters from the Council was an offer to sell G's council house to him but that they were merely invitations to treat which told him that the Council might be prepared to sell and of the possible terms of sale. So the application by G was not an acceptance of an offer to sell, but an offer to buy which the Council was entitled to reject.
>
> The courts' approach was defined as 'looking at the … documents relied upon as constituting the contract … (and) seeing whether on their true construction

there is to be found in them a contractual offer by the council to sell the house to Mr Gibson and an acceptance of that offer by Mr Gibson'.

Invitations to treat and offers in particular situations

Auction sales

An advertisement of an auction sale or a request for bids by the auctioneer is not an offer that the lot will be sold to the highest bidder, or to anyone else. It is merely an invitation to treat, inviting bidders to submit bids, which are offers to purchase the lot from the owner, the owner being represented by the auctioneer as his agent. The auctioneer then has the right to accept or reject the highest bid/offer.

Section 57 of the Sale of Goods Act provides that in an auction the sale is only complete when the auctioneer brings down the hammer. Until that time any bid may be withdrawn, and the auctioneer also has the right to withdraw a lot without accepting a bid. Each bid made during the auction lapses immediately a higher bid is made. Therefore, if a higher bid is made but is then withdrawn or rejected by the auctioneer, all previous bids will be void and the auctioneer must commence the sale again.

Where a lot is subject to a reserve price, if the auctioneer accepts a bid that is below that reserve price there is no sale, unless he has been given discretion to accept a lower bid. Where a lot is without reserve, if the auctioneer refuses to accept the highest bid there will not be a contract of sale between the bidder and the owner of the lot but the auctioneer may be in breach of an implied contract with the highest bidder that the sale will be without reserve and that therefore he has a right to have his bid accepted.

Shops and other retail environments

The display of goods in a market or shop window or on a shelf in a self-service store is not an offer to sell the goods but simply an invitation to any shopper to offer to buy those goods.

Fisher v. *Bell* (1961) B owned a shop which sold fishing tackle. He had on display in the window a knife, described as a 'Fisherman's Friend'. It was designed for use by anglers for extracting fishhooks and scraping, trimming and gutting fish, but it was an ejector knife, what is often called a flick-knife. It was, and still is, a criminal offence to offer to sell or to sell a flick-knife for possible use as a weapon. There was no argument that the knife, although intended for and being sold for a perfectly legitimate use, could be used as a weapon. If the display of the knife was an offer, then anyone could come into the shop and buy the knife by agreeing to pay the price. B was prosecuted by the local police for 'offering a flick-knife for sale'. B's defence was that the display of the knife was simply an invitation to treat, and that he therefore could refuse to sell if any unsuitable person entered his shop and asked to buy one. His defence was successful (and the law concerning the sale of flick-knives was amended as a result, to ban the 'display' of such knives).

Pharmaceutical Society of Great Britain v. *Boots Cash Chemists (Southern) Ltd* (1953) BCC wished to convert a number of its shops into self-service stores. The law is that controlled drugs, such as strong painkillers, can only be sold under the supervision of a suitable person such as a doctor or qualified pharmacist. BCC intended to have a pharmacist available at the cash-points in its stores to supervise sales if required. In a test case the Society prosecuted BCC for a breach of the law. Effectively the argument was similar to that in *Fisher* v. *Bell*. If displaying the goods on a supermarket shelf was an offer to sell the goods then a contract of sale might either be made when a customer took the goods from the shelf or when he took goods to a cash-point and offered to pay. If on the other hand displaying the goods was merely an invitation to treat, then the offer would take place when a customer took goods to a cash-point and asked to buy them. The contract would then be made when BCC accepted that offer, and this would be under proper supervision. The court accepted that for several, entirely practical, reasons the display of goods on a shelf in a supermarket is not an offer but an invitation to treat and that the sale contract only takes place when the customer asks to buy the goods. BCC was therefore not in breach of the law.

Offers to sell goods or services on a website

Offers on a website will usually be simply invitations to treat, because the seller will only have a finite number or quantity for sale and must therefore have the right to refuse to accept any further contracts once his supply has been exhausted. However, an offer on a website to sell software to be downloaded by the buyer on payment will be an offer, because the supply cannot be exhausted.

Advertisements and notices

The rules differ according to the situation. The basic rule is that an advertisement will only be an invitation to treat, leading probably to further discussions between the parties, or to one or more offers in response to the advertisement. See for example the case of *Partridge* v. *Crittenden* (1968) regarding an advertisement offering birds for sale.

The same is true of notices. For instance, a menu on display outside a restaurant or a price list outside a snack bar or theatre is an invitation to treat.

But there are exceptions –

Rewards and prizes

An advertisement or notice of a reward or a prize for doing something is usually an offer which leads to a unilateral contract. (A unilateral contract is a contract under which only one side has to do something. No one has to do whatever is necessary to earn the reward, but if someone who is aware of the offer does so then the offeror has to pay.) See the case of *Williams* v. *Carwardine* (1833). A reward was offered for information leading to the conviction of the murderer of Mr C. Then W, who was aware of the advertisement, gave information 'to ease her conscience', and claimed the reward. Payment was refused, but the court ruled that she was entitled to receive it. Also see the case of *Carlill* v. *Carbolic Smokeball Co* below.

Of course, the offer can be withdrawn at any time. But if the offeror knows that someone has commenced actual or potential acceptance, he cannot withdraw the offer from that person.

Vending machines

A notice displayed on or beside an automatic vending machine will, depending on the circumstances, either be an offer, because it will set out the terms on which the proprietor offers to sell items from the machine, or, more usually, the terms that will apply to the offer to buy those items made by anyone using the machine.

Ticket machines

An advertisement or notice leading to the direct or automated purchase/issue of a ticket may well be an offer. Automated machines that sell railway tickets are a typical example. Also, a notice displayed at the entrance to a car park setting out the terms upon which the public may use the car park will be an offer which is accepted every time someone drives in or takes a ticket; see *Thornton* v. *Shoe Lane Parking Ltd* (1971) Chapter 10.

Notices leading to a contract

The classic case concerned the hire of a deckchair.

> *Chapelton* v. *Barry UDC* (1940) C hired two deckchairs on the front in the town of Barry. A notice by the stack of chairs simply declared that they were for hire at two pence each. C took two chairs from the stack. A short time later he was approached by the deckchair attendant, duly paid his four pence and received a ticket. On the back of the ticket there was a clause that restricted the Council's liability for injury. When C opened one of the chairs and sat down on it the canvas ripped because it was rotten. C, who was an elderly man, fell through the chair, injuring his back quite severely upon the paving stones beneath. He claimed compensation. The Council refused his claim on the basis that the ticket was the contract document. The court rejected this argument. The clause could only be effective if it had been brought to C's attention at the time he made the contract. He made the contract by taking the chairs. The terms of the contract were those stated in the notice. The ticket was given to him simply as a receipt, and after the contract had been made.

European Union public procurement

Under the rules of the European Union relating to procurement by government and other regulated bodies, any significant purchase (above a basic value) has to be advertised in the *Official Journal* of the EU, and any organisation within the EU can then purchase the enquiry and bid for the resulting contract. If bona fide offers are made in accordance with the rules set out in the advertisement and the enquiry, the offeror submitting the lowest or best bid has a right to have his offer accepted. Currently the rules are in the process of being updated in accordance with three EU Directives, 2014/23, 2014/24 and 2014/25, published in February 2014 and due to be in operation sometime in 2016.

The law's approach to analysis

The approach is best illustrated by the instructive and entertaining case of the carbolic smokeball. This was a case of an advertisement for a 'reward', or rather for compensation described as a reward, but where the offeror actually refused to pay against a claim. It was also virtually the first consumer protection case.

Carlill v. *Carbolic Smokeball Co* (1893) The company inserted an advertisement in several newspapers – '£100 reward will be paid by the Carbolic Smoke Ball Company to any person who contracts the increasing epidemic of influenza, colds, or other disease caused by taking cold, after having used the ball three times daily for two weeks according to the printed directions supplied with each ball. £1,000 is deposited with the Alliance Bank, Regent Street, showing our sincerity in the matter'. (Remember that in 1891 £100 was equivalent to anything up to £100,000 today.) C read the advertisement and bought one of the balls. She used it as directed from 20 November 1891 until 17 January 1892, when she caught flu. She claimed the £100 reward. (Incidentally, the smokeball was a mixture of several noxious chemicals including coal tar, menthol, carbolic acid and sulphur. The method of use was to heat the ball and inhale the fumes for several minutes. To do this three times a day must have been fairly difficult. Sulphur fumes in particular are pretty unpleasant.) When the case came to court the company produced a whole series of defences.

The first was that the advertisement was 'mere puff' and that there was no intention that it should be taken seriously. The court disposed of this very easily. The simple fact that in its advertisement the company had referred to a substantial deposit of funds in a reputable bank was quite enough to signify to possible purchasers of the smokeball that the company intended the advertisement to be taken very seriously indeed. It could therefore be a legal offer.

Second, the company claimed that the offer was too vague because it did not state any period within which a user of the ball must catch influenza. The response of the court was that the defence failed because C was actually still using the ball when she caught the flu.

Next the company claimed that it was impossible to make an offer to 'all the world'. An offer must be made to an identified person. The court simply pointed out that this was a fallacy. It is completely correct to say that a contract must be between identifiable persons, but an offer may be made to anyone who may wish to accept.

The company then claimed that C had not told it that she had accepted the offer. The court responded that the company had not asked her to do so.

Finally, the company claimed that C had not given any consideration because C did not buy the ball from it. The court responded that C had given consideration. Although she had bought the ball in a shop, that ball had been manufactured by the company and had then been sold by the company through the retail chain. Therefore the company gained from their sale of the ball and that was consideration enough.

But in fact, was this so? The consideration was at best no more than something that might encourage the retailer to buy a further smokeball to replace the one sold. And was this consideration really necessary at all? If what was on offer was a reward, then catching flu was the consideration. But if what was on offer was really compensation, perhaps the court felt it necessary to put the consideration issue beyond doubt.

As a result the court ordered the company to pay £100 to C. The company did so and then immediately went into liquidation, presumably because there was a large array of other claimants queuing up to claim £100 as well.

5.6 Offers

An offer or provision of information?

In principle the difference between an offer and the provision of information is simple. Providing information is a normal part of the preliminary stages to a contract and cannot lead to any obligation, whereas an offer can do so if it is accepted. But information may be provided during the offer/acceptance process. And in practice, in some situations when providing information to the other side about the terms of a contract, it is easy to move from giving information to making an offer. Compare these two cases.

Harvey v. *Facey* (1893) H sent a telegraph message to F, who was on a long train journey, saying 'Will you sell us BHP (a farm)? Telegraph lowest cash price'. F telegraphed in reply 'Lowest price for BHP £900'. H telegraphed in reply that he accepted, 'I agree to buy BHP for the £900 asked by you'. The court held that in the circumstances F was only providing information as to the lowest price at which he might be prepared to sell rather than making an offer at that price (particularly since he had not answered 'Yes' to the question in H's first telegraph).

Bigg v. *Boyd Gibbins* (1971) After negotiations, BG wrote to B rejecting an offer for a property at a price of £20,000, but then added 'for a quick sale I would accept £26,000 …' B accepted this price and BG then acknowledged B's acceptance and stated that he was passing the matter to his solicitor so that the solicitor could proceed with the sale. The court held that in this situation BG had not merely provided information but was also giving notice to B that he was prepared to sell at that price provided that B accepted immediately. There was therefore a valid contract for the sale of the property.

The budget offer

A budget offer is an incomplete or part offer, or an offer subject to further negotiation or clarification, in other words, a submission to the offeree which gives him a considerable amount of information about what the offeror is prepared to propose to him. It is not an offer, though it is recognised as an important part of commercial negotiations. It is actually an invitation to treat, as it signifies willingness to continue the discussions should the offeree find the proposal of interest.

What constitutes an offer?

An offer is a promise of a contract on the specific terms set out and referred to in the offer if those terms are accepted in full and without qualification. That acceptance must be made by the offeree, or an offeree to whom the offer has been made, and before the offer expires.

The rules are simple.

The offer must be complete. It must include, actually or by reference, the full terms of the potential contract, in detail. If the terms are vague, unclear or uncertain then the document is not an offer that can form the basis of a contract. An example of this is the case of *Scammell* v. *Ouston* (1941) referred to in Chapter 4 above.

> Another is *Gunthing* v. *Lynn* (1831) A purchaser offered to buy a horse on the basis of a basic price, plus a further sum if the horse, a racehorse, turned out to be 'lucky'. The owner accepted the offer, but then decided that he did not wish to sell and the court decided that the price stated in the 'offer' was too vague for there to be a contract.
>
> A different example of the same principle – a 'referential' offer – is the case of *Harvela Investments Ltd* v. *Royal Trust of Canada Ltd* (1986). This case concerned an offer for the sale of a large parcel of shares in a company. The prospectus/enquiry requested fixed price bids. One bidder offered to buy the shares for £xx,ooo more than any other bidder. The court decided that this bid could not qualify as a fixed price.

But where possible the law will interpret the language used by the parties in a way that will give a commercial agreement contract status. See the cases of *Hillas & Co Ltd* v. *Arcos Ltd* (1933) and *Nicolene Ltd* v. *Simmonds* (1953) referred to in Chapter 4.

Of course, if the language used cannot be interpreted clearly then there is no offer. In the case of *Falck* v. *Williams* (1900) an offer was made by telegram in a compressed form in broker's code. Unfortunately, the message was so compressed that an essential full stop was left out. It could have been placed before one particular word or after it. The result was that the message had two different possible meanings. So an attempt to accept the offer did not create a contract. Compare the case of *Lind (Peter) & Co* v. *Mersey Docks & Harbour Board* (1972) below.

An offer can be made by actions, though this is rare in all but the simplest of transactions. Perhaps an example would be flowers or produce placed by a garden gate with a notice stating 'xx pence per bunch, put money through letterbox'. However, almost all offers, and certainly all offers in the commercial world, are made in words. The offer may be made verbally or in writing, and by any method of communication.

The offer must be communicated. This is so obvious that it hardly needs to be stated, but it can be important in a 'reward' situation. If A offered a reward for the return of a lost dog and B found and returned the dog without being aware of the offer until afterwards, could B then claim the reward? Legally the answer has to be in the negative, though in 99% of such cases A would make payment almost automatically.

Most offers made in a commercial context will be subject to a fixed validity period.

If an offer is not subject to a fixed validity period, it will terminate at the end of a reasonable period. What is a reasonable period will depend upon the circumstances of the case. In theory this is satisfactory, but there is no real guidance on what a reasonable period is. The only case that has ever dealt with the question related to an offer to purchase shares and, given the actual delay that occurred, the case is not very helpful when dealing with normal commercial contracts.

> *Ramsgate Victoria Hotel Co* v. *Montefiore* (1866) M had offered to buy shares in R in June and had paid a nominal deposit. He heard nothing more until late November when he was informed that the shares had been allotted to him and that the balance of the price for the shares was now payable. In the meantime M had invested his money elsewhere. The court held that it was unreasonable to expect that the offer to buy the shares made by M in June should still be valid five months later. So M was not obliged to stand by his offer. But the court did not offer any guide on how to decide what a reasonable validity period might be.

(The case was discussed later in *Manchester Diocesan Council for Education* v. *Commercial and General Investments Ltd* (1970) but again without any proper guidance being offered.)

Of course, in a commercial relationship the problem of a reasonable period may not too hard to solve most of the time. Most significant offers will be made subject to a fixed validity period. The parties will often have had a history of previous dealings which will suggest a probable validity period, or there will be a custom in the trade which will suggest an expected period. However, there must always be some degree of uncertainty in any offer situation as to what the practical effects of the 'reasonable period' will be. This uncertainty will be at its greatest in a novel relationship with no previous history of dealings between the parties.

So it is always advisable to give every offer a fixed validity period.

An offer may be subject to a condition

There are two types of condition, precedent and subsequent.

An offer subject to a condition precedent only becomes effective if the condition is met. A typical example might be 'If our suppliers confirm that they can fit your requirement into their next production run, we can offer you a delivery date within the next four weeks'.

If an offer is subject to a condition subsequent then the offer will lapse if it becomes impossible to satisfy that condition. For instance, a seller might offer to supply products by a date, 'subject to shipping space still being available at the time of order'. If at that time space is not available, the offer would lapse. Alternatively the offer might be deemed to consist of two separate offers, one to supply by the stated date if space was available, plus a second offer to supply by a reasonable date if it was not; see the section on 'Timing' below.

A condition may be stated in the offer or may be implied.

Financings Ltd v. *Stimson* (1962) On *16 March* S purchased a car from X, a dealer, and signed an application for a hire-purchase contract with F. The application was in the form of a standard agreement between S and F and included the following, '[t]his agreement shall be binding on F only upon signature on behalf of F'. On *18 March* S paid the deposit and took delivery of the car. On *20 March* S returned the car to X because he was dissatisfied with it and agreed to forfeit his deposit. On *24 March* the car was stolen from X's premises and was badly damaged. On *25 March*, in ignorance of all this, F signed the agreement. Later F sold the car for what it would fetch and sued S for payment of the balance of the hire-purchase price.

The court dismissed F's claim. The agreement signed by S was an offer to enter into an agreement with F. That offer must be subject to the implied condition that there was no material change in circumstances before its acceptance by F and in particular that the car would still be in substantially the same state as it was when it was purchased by S. Also, X was acting as an agent for F. So when S returned the car to X he revoked his offer to F, and this revocation was accepted by X as an agent for F.

5.7 Post-offer situations

Revocation

Under English law an offer may be revoked, or withdrawn, by the offeror at any time before it is accepted. (This right of revocation is very much a peculiarity of English law. It is not permitted under continental systems, nor is it permitted under Scottish law or most other systems derived from common law.)

But if the offer is made as a deed or under seal it is irrevocable and must remain open for its validity period. If an offer is made for consideration, 'In return for the payment of £xx it is agreed that the offer shall remain open for acceptance until (a date)', then that offer can be revoked but revocation would create liability for damages for breach of the contract to keep the offer open.

Revocation must be communicated to the other side for it to become effective.

> *Byrne* v. *Van Tienhoven* (1880) VT was based in Cardiff, B in New York. On *1 October* VT wrote to B offering a quantity of tinplate for sale. On *8 October* VT posted a second letter to B revoking the offer. On *11 October* B received VT's first letter and immediately sent a telegram notifying his acceptance of the offer which he then confirmed in a letter posted on *15 October*. On *20 October* VT's second letter arrived in New York (and on *22 October* B's formal letter of acceptance arrived in the UK). Of course this case relates to a time when communication between the UK and America was very different to what it is today, but that only makes the principles more clear. The court held that the revocation letter did not become effective until *20 October* when it arrived in New York. Therefore the contract had already been made before the revocation letter arrived and the revocation letter was ineffective.

What is clear from this decision is that if anyone does need to revoke an offer then it is vital to tell to the other side as quickly as possible. There is always the risk that the other side might accept the offer before it is revoked. If in *Byrne*'s case VT had not sent a letter but instead had telegraphed his revocation on 8 October (in the Victorian era the telegram was the equivalent of a telephone message or email today), the problem would not have arisen because notice of revocation would have reached America even before the offer.

It is not necessary for information that an offer has been revoked to be given by the offeror, provided that information is given by a reliable person or comes from a reliable source.

> The difficult case of *Dickinson* v. *Dodds* (1876) is relevant. Dodds wrote to Dickinson on 10 June offering to sell a house to him, saying that the offer would be open until 9 am on the 12 June. On 11 June Dodds sold the house to A. That evening

Dickinson was told this by B, who was not in any way connected to Dodds. Dickinson then left a written acceptance of the offer at Dodds' address the same evening and requested B to give Dodds a copy, which he did before 9 am the next morning (the twelfth). Dickinson later sued Dodds for failure to transfer the property to him. (The facts of course appear to tell a tale of complete incompetence, and worse, which perhaps affected the decision of the court.) The court rejected Dickinson's claim on the basis that Dickinson knew that Dodds no longer intended to sell the house to him as clearly as if Dodds had told him so himself. As Dickinson knew that the offer had been withdrawn he could not accept it.

The court clearly considered that Dickinson was required to rely on what he was told by B. (The court obviously considered that B was a reliable witness, and that Dickinson thought so too.) But what would the situation have been if B regularly got his facts hopelessly wrong (and the timescale was such that it was probably almost impossible for Dickinson to verify the facts before 9 am next morning anyway)? There is only one other decided case on the point, which does not actually help. The only thing that can be said is that each case would be decided on its facts.

There is another problem too. *Dickinson*'s case referred to the sale of a specific item, a house. But imagine a different situation, relating to generic goods. A manufacturer submits offers to two different customers, A and B. Each offer is for a large quantity of equipment which would effectively absorb most of his manufacturing capacity for the next six months. The offers both quote a delivery period of six months from date of order. What is the position for A if he is told by a third party that B has accepted when A is also about to accept? Does A need to confirm with the manufacturer that the offer is still open before accepting it? After all, in the commercial world it is always possible to subcontract the work. Even worse is the situation where an article appears in the press announcing B's contract, whether or not the article is seen by A. (Of course in practice the manufacturer should deal with the situation properly by informing A immediately whether or not his offer was still open.)

Acceptance

A judge speaking in the early years of the twentieth century, when a promise to marry was enforceable as a contract, explained the various methods of acceptance as follows. If a man asks a lady to marry him, but only succeeds in producing an embarrassed mumbling, so that she cannot understand, then no offer has been made. If when he succeeds in asking properly the lady replies with a scornful laugh or a vigorous shake of the head, then she will have rejected his offer by her actions, because these are actions which would be interpreted by any reasonable person as a rejection of the proposal. If she says 'No' or 'You are joking, of course', in a clear voice, then she will have rejected his offer by her words, directly or by implication. If she says 'Yes', or 'Where shall we go on our honeymoon?', then she will have accepted his offer in words, because they accept the proposal, again either directly or by implication. If, however, as she replies 'Yes', her voice is drowned by the sound of a passing train, so that the man cannot hear what she has said, then no acceptance will have taken place because that acceptance has not been properly communicated (unless of course the man can lip-read), but the offer will still remain open. Finally, if she says nothing but throws herself into his arms then she will have accepted his offer by her action, because this is an action that would be interpreted by any reasonable person as an acceptance of his proposal.

An offer can only be accepted by the offeree, or one of the offerees, to whom it was made, or by his or their agent(s). Normally acceptance must be communicated by the offeree to the offeror, but see *Carlill*'s case above and *Felthouse* v. *Bindley* below. Communication of acceptance by a third party is not enough to create the contract.

Acceptance must be intended. An 'acceptance' of an offer of which the offeree was not aware does not make a contract. Acceptance must also be clear. This is the result of a case where an offer included different options.

Lind (Peter) & Co v. *Mersey Docks & Harbour Board* (1972) PL bid to MDHB for a contract to build a mole, offering two pricing options: either to do the work on a rates basis or for a fixed price (in the region of £14 million). MDHB accepted PL's bid by simply saying 'we accept your offer'. PL carried out the work, but the project turned out to be difficult and cost much more to complete than the fixed price that PL had quoted. PL invoiced for the full cost of the work, plus profit, and MDHB paid only the quoted fixed price. PL sued for the unpaid amount. The court decided that PL had not made one offer, but two, the first at a fixed price, and the second based on rates. MDHB had not said which offer it accepted, so it had accepted neither. The result was that there was no contract, and PL was entitled to charge for its work on a *quantum meruit* basis of cost plus a reasonable profit.

An offer can only be accepted within its validity period.

Acceptance can be by words or by actions. Acceptance in words will be any statement that directly accepts the offer, in full and without any qualification, or any statement that implies acceptance. Acceptance by action will be any deliberate act or conduct that implies acceptance of the offer. An acceptance of an offer in full but with the addition of an additional term is not an offer but a counter-offer; see *Northland Airliners Ltd* v. *Dennis Ferranti Meters Ltd* (1970).

Once an offer has been accepted, a contract will come into being. The date/time of the contract will be the moment of acceptance. The terms of the contract will be those of the offer.

We have already seen in *Carlill* v. *Carbolic Smokeball Co*, above, that in a unilateral contract the person offering the reward can dispense with the need for anyone commencing performance to give notice that he may accept the offer by claiming the reward. In all normal situations, however, acceptance must be positively communicated to the other side.

> *Felthouse* v. *Bindley* (1863) Uncle F wrote to nephew F offering to buy a horse that was to be auctioned for £30 15s, adding 'If I hear no more about him I (will) consider the horse mine at that price'. Nephew F certainly intended to accept the offer, and as a result asked the auctioneer B to withdraw the horse from the auction. Due to an administrative mistake, B sold the horse to a third party and uncle F sued B for selling his(?) horse. The court rejected the claim, on the basis that this would allow anyone to impose a contract upon another by claiming that there would be a contract unless the other side took positive action to refuse his agreement.

And in a much more recent case, *Allied Marine Transport Ltd* v. *Vale do Rio Doce Navegacao SA* (1984), the court said that silence is simply equivocal. It could mean anything, acceptance, forgetfulness, or just hoping that an issue will not require a decision.

In *GHSP* v. *AB Electronics* (2010) we saw that an action that could be construed as acceptance following a series of offers would create the contract. Two more examples of the same principle –

> In *Nissan (UK) Ltd* v. *Nissan Motor Manufacturing (UK) Ltd* (1994) the court said 'If two parties to-and-fro with offers and counter-offers, and one maintains a proposal to the last with no come-back from the other, it could naturally be inferred that any subsequent conduct by the other party that was referable to the existence of some contract between the parties signified acceptance of that proposal'.

> *Tekdata Interconnections Ltd* v. *Amphenol Ltd* (2010) was a case which resulted from a long-term course of dealing. The commencement of the relationship was a purchase order from the buyer TIL, and in response AL quoted its own sale conditions. There was no other 'contractual' correspondence, but the parties then dealt with each other for several years, and several purchase orders, until there was a dispute. The court held that the final contract in the series was still governed by the conditions that applied to the initial contract. So AL's conditions applied.

Non-acceptance

If an offer is not accepted during its validity period then it will terminate.

Rejection

If an offer is rejected by the offeree then it will terminate immediately.

Rejection must be communicated to the offeror. It may be by words or actions. Words will comprise any statement that directly or by implication rejects the offer. Action will be any act that implies rejection of the offer. What those words or actions must be will of course depend upon the circumstances.

Lapse

All offers will lapse/terminate at the end of their validity period.

Counter-offers

A counter-offer will reject the original offer, but replace it by a further offer to the previous offeror.

The simplest example of rejection by counter-offer is the following.

> *Hyde* v. *Wrench* (1840) On *6 June* W offered to sell a property to H for £1000. On *8 June* H made an offer of £950. On *27 June* this offer was rejected by W. Then on *29 June* H wrote to W saying that he now accepted the offer of 6 June. In the meantime W had decided that he no longer wished to sell. H sued for breach of contract, but the court decided that no contract had been made. By making a counter-offer H had signified that he rejected the original offer made by W, and he could no longer accept an offer that he had rejected.

5.8 Communication rules

The rules about the communication of offers and acceptances, etc., are as follows.

The postal rules –

A letter accepting an offer takes effect from the moment that it is *properly* posted, which means posted into a posting box or handed to an employee of the post office who is authorised to accept it.

A telegram accepting an offer also takes effect from the moment it is sent.

A letter or telegram revoking an offer takes effect only from the moment that it is properly delivered.

(The references to telegrams are included because of the case of *Stevenson* v. *McLean* referred to below.)

All other letters take effect from the moment from the moment that they are received.

Other rules –

All electronic communications – facsimile, email, telex, etc. – are effective when received, and in the form in which they are received, so that any errors in transmission are the responsibility/risk of the sender.

Spoken communications are effective when heard and as heard.

Every person is responsible for his own communications. If he talks to someone else he is responsible for ensuring that the other party has heard him correctly and understands what he has said. If he sends a letter, email or any other communication, he is responsible for sending it to the correct address, and for what is then read by the recipient. In other words, the sender has the right to select the method by which he transmits the message, but he also must accept the risk that the message might be late or might arrive in a corrupted form. (The classic story of the message 'Send reinforcements I'm going to advance' being corrupted when passed by word of mouth into 'Send three and fourpence I'm going to a dance' may not be so funny if it happens in a commercial context.)

As an example of the communication risk of using a letter, see the case of *Adams* v. *Lindsell* (1818). L sent an offer to A by letter, insisting on a response by return of post (that is, immediately). Unfortunately, the letter was wrongly addressed, so that it arrived some days late. A wrote back immediately, accepting the offer, but L had already sold the goods to someone else. A sued for breach of contract and succeeded. The court held that as the delay was entirely due to L's mistake he must take the risk.

For a practical example of problems of communications in general see the case of *Byrne* v. *Van Tienhoven* referred to above.

Another example is the situation that arose in *Entores* v. *Miles Far East Corporation* (1955) relating to a telex communication. The facts were that an offer made by an Austrian company was accepted by a telex sent from London to Vienna. The question was, simply, was the resulting contract subject to English or Austrian law. The decision was that as the telex was effective only from its receipt in Vienna, the contract was made in Austria and was therefore subject to Austrian law.

5.9 Some more practical problems

Death

In general terms the death of the offeror or offeree will prevent there being any contract.

If it is the offeree who has died, then the offer will simply lapse because no one else will have the ability to accept the offer.

If it is the offeror who has died, then the offer will usually be terminated by his death. This is particularly true if his death has been notified to the other party and it is also true when the contract would require personal work or services to be provided by him, as for example would be the case in a contract of employment or for consultancy work.

However, it is not true in all circumstances. There is a small number of cases in which an agent has made or accepted an offer before being informed of the death of his principal. The cases conflict, and the only comment that can be made is that any such situation will depend upon the precise language used in the offer that is the subject of the dispute.

In addition, there is a line of cases concerning guarantees given to support bank overdrafts by friends or family members. The problem is that a guarantee of another's overdraft is effectively a standing offer which is taken up or accepted by the bank every time the holder of the bank account is allowed to overdraw. The law seems to be that if after the death of the guarantor, but before the bank has been notified, it allows the account holder to overdraw or makes a claim under the guarantee, the guarantee will still be capable of being enforced against the estate of the deceased.

Timing

An offer will not be terminated if the parties exchange information or ask questions concerning the details of that offer or continue their discussions. This is usual in the case of complex commercial situations anyway. However, it will terminate if it is rejected. It may be rejected in two ways – either by rejection or by counter-offer. To state the principle is easy, but the practical problem is that at times it may be difficult to distinguish between them. Consider the following case, which, like *Byrne* v. *Van Tienhoeven*, is an example of the problems caused by the difficulties of communication in Victorian times. Effectively this is an example of a situation where delay in communications meant that one of the parties had to lose, even though both had behaved perfectly properly.

> *Stevenson* v. *McLean* (1880) M was a merchant who dealt in iron and steel. He was in negotiation with S (amongst others). On Saturday M wrote to S, 'I would now sell [approximately 4000 tonnes of iron] for 40 shillings net cash [per ton] open till Monday for immediate delivery'. On Monday S sent a telegram to M, 'Please wire whether you would accept 40 [shillings] for delivery over two months, or if not, longest limit you would give … ' [for delivery and payment]. This telegram was received by M at 10.01. M then sold the iron to a third party. At 13.25 M sent a telegram to S stating that he had sold the goods to someone else. At 13.34 S sent a second telegram accepting the offer to sell at 40 shillings cash. At 13.46 M's telegram of 13.25 reached S. S's telegram accepting the offer arrived a few minutes later.
>
> The problem was quite simple. M's telegram of 13.25 was notification that he considered that he was in a position to revoke his offer and was therefore doing so. But this attempt at revocation did not arrive until 13.46. If, therefore, the offer to sell was still open for acceptance until that point, there had to be a contract. The only thing that could prevent there being a contract was if the first telegram from S on Monday had been a counter-offer which had terminated the original offer from M at 10.01.
>
> The case therefore turned on the nature of the Monday morning telegram from S concerning the delivery/payment period. Clearly M had interpreted this telex as being a counter-offer. The court rejected this view. The judge said that what

the telegram did was to ask a question, 'would you accept a delivery over two months, or if you won't accept two months how long would you accept?' Effectively therefore the telegram could not be a counter-offer, because it did not make any effective proposal of its own, but simply asked M for information about the degree of flexibility that there might be in M's position. The offer was therefore still open and was accepted by S's telegram of 13.34. The judge disposed of the telegram from M sent at 13.25 simply by saying that it could not be effective until it arrived. The only comment that one can make is that sometimes a telephone can be a blessing.

5.10 Contracts outside the offer/acceptance process

From time to time the courts have bemoaned the fact that the rules of the offer/acceptance process are followed too literally and have expressed the wish for a more flexible approach. But in practice the number of cases in which the courts have actually dared to adopt a more flexible approach is minute, and the courts have only done so in order to try to achieve a fair result, or, more correctly, to try to avoid a totally unfair result. None of these cases relate to anything like the normal commercial contract. The classic example is that of a race.

> *Clarke* v. *Dunraven* (1897) D and C both entered their racing yachts for a race organised by a sailing club. Entrants had to agree to abide by the club's rules for the race which included a provision that anyone damaging another yacht in breach of the race rules should pay the full costs of any damage that he caused. (At that time the normal rule under the Merchant Shipping Acts was that liability was limited to £8 per tonne, which was hopelessly inadequate for a racing yacht.) C's yacht negligently collided with and sank D's yacht. The court held that by agreeing to abide by the rules of the sailing club, C and D had entered into a contract with each other (and presumably with all the other entrants as well) because everyone who had entered knew that everyone else had agreed to abide by those rules. Therefore C was liable for the full value of D's yacht.

The case is difficult. It is a simple, fair and common sense decision. But where is the offer and where is the acceptance? The court did not explain, probably because it did not want to. The only possible way that we can see is to say that when the first entrant agreed to accept the sailing club rules, he made an offer to abide by those rules to everyone else who might enter. When the second entrant agreed to accept he accepted the offer made by the first entrant, and also made an offer to everyone else who might enter, and so on. This looks attractive at first sight but does not sit comfortably within the rules relating to the offer, or to the acceptance.

The standing offer

There is one kind of offer that needs particular mention. Most offers are intended to form the basis of a single transaction. Some offers are, however, intended to form the basis of multiple transactions, sometimes over a considerable timescale. This is called a standing offer.

> *Great Northern Railway Co* v. *Witham* (1873) GNR advertised for offers for various categories of items. W submitted an offer in GNR's prescribed form, which included the following: 'I undertake to supply GNR for twelve months with such quantities of (the items) as GNR may order from time to time'. GNR replied, accepting the offer from W, and subsequently placed a number of orders with W which were carried out in normal fashion. Finally, GNR placed an order for items, which W was not prepared to meet. W therefore revoked his offer and refused to carry out the order from GNR. GNR sued W for breach of contract and was successful. The court held that the offer made by W was a standing offer that was capable of forming the basis of a series of contracts, as often as GNR accepted by placing an order for a quantity of the items in question. Therefore, although W could withdraw his standing offer at any time, any order placed by GNR before the offer was withdrawn was a valid contract.

The casual approach

As is so often the case, one of the clearest examples of this is one of the nineteenth-century cases.

> *Brogden* v. *Metropolitan Railway Co.* (1877) B was a coal merchant who regularly supplied coal to M. After some years the two decided to formalise their relations. M therefore sent a draft of a contract to B setting out the terms on which M would purchase coal in future. B completed a blank space for the name of an arbitrator in the 'disputes' clause, signed it and returned the draft to M marked 'approved'. M's buyer did not sign the document, but simply put it into a drawer. There was no further communication between the parties about the contract, but the two parties then continued to trade as before but on the terms set out in the draft. Later there was a dispute and B refused to supply any longer, claiming that no contract had been made between them. The court decided that there was a contract between them, and that therefore B was in breach.

However, the court was not very sure about precisely when that contract was made, in other words, which action by which party actually created the contract. Clearly when the buyer sent the draft document to B he made an offer on behalf of M. Equally clearly, when B inserted the name of an arbitrator and returned the draft, he made a counter-offer to M. M did not respond to the offer by B at all, but the subsequent conduct of the parties could only be explained if the terms of the draft had been agreed to be a contract. The court was divided:

- either the contract was made when M first ordered coal after the draft was filed away in the drawer (in other words, by its action in ordering coal M accepted the 'offer to sell' made by B in returning the draft contract) or
- the contract was made when B delivered coal against that order (in other words, M was making an 'offer to buy' on those same terms which was then accepted by B).

Of course, in the circumstances the difference between the two made no difference in practice, but the case is a perfect example of how poor communication or administration procedures may cause legal uncertainty.

Legal uncertainty in a dispute is always very expensive.

Conditional acceptance

A 'conditional' acceptance of an offer is not acceptance. If it is a partial acceptance it will be a counter-offer as we have already seen. If it is an acceptance 'subject to contract', to use the phrase used so regularly in connection with the sale of land or buildings, then it is simply notification that the party is ready to accept the offer as soon as the other side is willing to enter into a contract in the appropriate form. Sometimes there can be difficulties in deciding the exact status of an 'acceptance' or 'agreement'.

> *Branca* v. *Cobarro* (1947) The parties entered into a written agreement for the sale of a small business. The agreement stated that it was 'a provisional agreement until a fully legalised agreement, drawn up by a solicitor and embodying all the conditions herewith stated is signed'. The court decided that the agreement constituted a full and therefore enforceable contract that was simply intended to be replaced by a more formal contract in due course.

However, there are two very important aspects to the case. First, the dispute went as far as the Court of Appeal, a long and expensive dispute about something that should not have been a problem. The second aspect was the court's reasoning. The 'provisional' agreement was to be upheld because it contained everything necessary for a contract. Therefore it was a contract, whatever it called itself. If on the other hand the parties had used clear language to declare that the agreement was not to be enforceable, the decision would have been different. Interestingly the court accepted that the phrase 'subject to contract' would have been adequate to achieve this. (And see also Chapter 15 below.)

Silence is not consent except in unusual circumstances

We have already seen in *Carlill* v. *Carbolic Smokeball Co* above that in a unilateral contract the offeror of a reward can dispense with the need for anyone commencing performance to give notice that he may accept the offer by claiming the reward. In all normal situations, however, acceptance must be communicated to the other side. 'Silence is not consent'. There are only a very few possible exceptions to this.

An example is where the parties have already agreed to accept a contract without the need for any formal offer/acceptance process. A typical example of this is that of goods held on a 'sale or return' basis. If a bookshop has taken delivery of books on a sale or return basis, then each time the shop sells one of the books a contract will come into being between the shop and the supplier for the purchase of that book by the shop, even though there has been no communication between the shop and the supplier. (Otherwise the shop could have no right to sell the book.)

Finally

In the commercial world, of course, many contracts are made in two completely different ways that do not properly conform to the traditional view of offer and acceptance.

First, many significant contracts are made as the result of a series of negotiations between the parties. In these cases there may well be an enquiry/invitation to treat, followed by an offer, or a series of offers, but then followed by a negotiation process to agree the final terms of the deal. The original offer/acceptance process will now be superseded by an offer and acceptance based upon the terms as finally agreed, with the parties in effect saying to each other 'I will accept this if you will'.

Second, many contracts result from the 'battle of the forms', a series of invitations to treat, offers and counter-offers between the parties, each based upon that party's own in-house terms. The case of *GHSP* v. *AB Electronics* (2010) referred to above is one example. Another, the classic case, is *Butler Machine Tool Co Ltd* v. *Ex-Cell-O Corporation (England) Ltd* (1979). The case concerned a contract for the supply and installation of a milling machine by BMT for ECC. BMT quoted a price of £75,000, plus the cost of installation and commissioning. ECC sent an order to BMT at a price of £75,500 including installation/commissioning. The order included a tear-off slip to be signed/dated confirming acceptance of the order. BMT completed and returned the slip with a covering letter asking for the terms of their original offer to be reinstated. No further exchanges took place. The court held that the decisive document/event was the return of the slip. This completed the acceptance of the order even though BMT was still making it quite clear that it wanted an increase to the price of £75,500. So BMT failed in a claim for the extra sum.

The basic rule for any commercial organisation is very simple. Do not do anything that can be construed as agreement to a contract until you are satisfied that the terms are acceptable to you.

6

Words in Contracts Part 1 – Words Used Pre-Contract

The successful offer/counter-offer becomes the contract. So the words in that offer matter. To understand what words might be included in the offer and what they will mean, we need to know the answers to a number of questions.

- What did the parties actually say (and write, email, etc.) to each other?
- Which of the things that they have said will become a part of the offer?
- Are there any other things that they have said that will affect the offer/contract?
- What other terms might be implied into the contract as well as what is said in the offer?
- What importance will the different terms of the contract have?
- What is the effect of public policy on the contract (such matters as unfair contract terms and so on)?
- How does the law, or rather, how do the lawyers, interpret the words?

   This chapter is concerned with the first three of these questions. The others are dealt with in the next and succeeding chapters.

   The law always takes the simple approach. In the commercial world everyone will have heard and remember what has been said to them. Everyone will have read and will remember everything that has been given to them to read. The practice can be very different. We may not hear what has been said to us, especially if we are hard of hearing, or the room is too noisy, or if we are distracted at the critical moment. We may not read what has been given or shown to us. We may indeed be unable to read at all, or unable to read because we have forgotten to bring our reading glasses. Then we may forget what we have

Using Commercial Contracts: a practical guide for engineers and project managers, First Edition.
David Wright.
© 2016 John Wiley & Sons, Ltd. Published 2016 by John Wiley & Sons, Ltd.

been told, or remember only part of what we have been told. Something said to an engineer may not be known to anyone in the purchasing department or vice versa. Finally, memory can be a notoriously fragile and selective instrument at the best of times.

The problem of perception/retention of information is then made more complex by the problem of proof. If challenged, could we prove exactly what information we gave to someone else, and to whom we gave it and when? And what information they gave to us? And then there is the vexed question of whether that person properly understood the information we gave him. Of course, all this is not strictly a question of contract law at all. It is really a question of evidence. But it is always important in practice to be able to produce some evidence of what has been said by one party to the other during the pre-contract discussions.

6.1 The different types of statement

Pre-contract discussions can take place for many hours spread over a period of weeks or months. They will include different types of information.

First of all there will be 'advertising puff'. Every seller is automatically expected to talk up his product or services, telling the buyer how wonderful, marvellous, reliable or perfect it or he is. This is a recognised part of the sales/marketing process, and every commercial buyer is expected to be able to recognise and discount it.

Then there are the normal exchanges of information between the two sides that take place during many contract negotiations, for instance, when the experts discuss requirements and capabilities. Both sides are simply verifying their understanding of what the other side needs and can offer.

Next there will be what the law calls 'representations', statements made by one party to the other which may not become part of the offer but which will be made with the intention to persuade, and which may persuade, that other party to accept the offer when it is made.

Finally, there will be statements made during the discussions which will become a part of the offer, and in due course part of the terms of the contract.

Of course it is easy to define these different types of statement in a book. In real life it may be very difficult to distinguish one from another, and there are many cases in which statements made by one party to the other have been the subject of considerable discussion. For instance, the statement, 'we will always have a full range of spare parts in stock', could be advertising puff, or part of the normal exchange of information. It could also be a representation or even become a term of the contract.

Advertising puff

Advertising puff is a statement that is a part of the game but not really meant to be taken seriously, and one that no reasonable person would take seriously if he thought about it. An example of this was –

> *Lambert* v. *Lewis* (1981) A manufacturer's brochure said that one of his products, a tow-bar, was 'foolproof' and 'required no maintenance'. But nothing is ever totally proof against an idiot, or is guaranteed never to need any attention throughout its working life. A purchaser bought one from a garage. It later failed because an important part had been broken/lost (by the actions of the purchaser, though not by his fault) and caused a serious road accident when a trailer became detached while being towed along a main road. The purchaser, or rather his insurers, claimed that the effect of the statements in the brochure was to create a separate contract between him and the manufacturer, collateral to his contract with the garage, and that therefore the manufacturer should be liable for the accident. The court held that the manufacturer had no liability to the purchaser. No reasonably sensible person would have believed that the statements in the brochure were ever intended to be a guarantee of the product, so there could be no collateral contract.

But what is advertising puff may well differ according to the circumstances. A statement made by one commercial organisation to another as a part of the commercial game might well be puff. The same statement made by a commercial organisation to a non-professional non-expert consumer might be much more than that.

Good examples of puff can be found in the second-hand car trade, when salesmen will make statements such as 'a good runner', 'a superb car', and so on. But we can all see how advertising puff can turn into something much more substantial. If a salesman says that a rusty elderly banger 'is in perfect condition', that is no more than puff. But that puff does also imply a second statement – that the car is at least in reasonable working order, so that if the brakes fail or the engine falls out within a few miles after being driven away by a buyer, the dealer would be liable.

Exchanges of information

A buyer may want to check the supplier's ability to manufacture a product or whether he has the manufacturing capacity to be able to manufacture that product in the quantity or within the timescale that he requires. The supplier may want to check what the buyer's precise requirements are or whether he has the ability to pay.

Statements intended to persuade or representations

A representation is a statement of fact made by one party to the other with the intention of persuading him to enter into the contract. When that statement is correct there is nothing wrong. But if the statement is incorrect then it is a misrepresentation. If it then does persuade the other party to make the contract, there is a problem.

> *Yam Seng Pte Ltd* v. *International Trade Corporation* (2013) During negotiations leading up to a distribution agreement for toiletry products in the Far East, ITC told YS that it had already signed the necessary licences to enable it to manufacture and sell the products. The statement was untrue and ITC knew that it was. (One licence was actually signed only a few days before the distribution agreement, and the other not until several months later.) The result was that YS was persuaded to enter into the agreement several months earlier than might otherwise have been the case. There were then significant delays by ITC in bringing the products to market, which caused significant losses to YS. The court held that YS were entitled to damages for the misrepresentation.
>
> *East* v. *Maurer* (1991) During negotiation for the sale of a hairdresser's shop, the seller told the buyer that another shop that he also owned nearby would close. The buyer bought the shop. The seller did not then close his other business, so that the buyer lost trade. The buyer was awarded damages.

The basic principle is 'I would not have made that contract at that time if I had known that the position was different'.

The terms offered

Finally, there will be statements made by one party to the other as a specific part of the offer, or as a part of the negotiations leading up to the offer, on the clear understanding that they will become a part of that offer, when it is finally made.

Disputed terms

This will normally happen post-contract. In other words, an offer will have been made and accepted. The question will then arise whether or not a particular statement made before the offer but not specifically included within it should be considered a part of the contract.

Of course, complex commercial offers will usually be made entirely in writing with all the applicable terms set out clearly. Offers will often exclude previous discussions, to avoid problems of this kind.

But many simpler commercial contracts are made on the basis of very simple written offers, or offers that are a combination of oral and written terms, and the vast majority of consumer contracts are made purely on the basis of oral discussions.

So when can an oral statement made during discussions leading up to an offer be a part of that offer?

The law works on the basis of 'presumed intention' – whether a reasonable person would consider that it was the intention of the parties that that particular statement should be incorporated into the offer or not. There might be two reasons why this should be the case: that the statement is of great importance, and that one party is relying on the other's knowledge and/or expertise.

Importance and knowledge –

Bannerman v. *White* (1861) W was a corn merchant; B was a brewer. W had a quantity of hops for sale. B wanted to buy hops for use in flavouring his beer. Hops are vulnerable to fungus, and at that time a standard way of protecting them was to spray them with sulphur. Treated hops then had to be washed before they could be used, which much reduced the quality. B therefore asked W whether or not his hops had been treated with sulphur. He said that he was simply not interested in buying them if they had. W, who knew exactly why B asked the question, replied with complete honesty that the hops had not been treated. B therefore bought them. The sale note/contract made no mention of the point. Unbeknown to W, some of the hops had been sprayed with sulphur. (The grower had simply lied to W to get a better price.) When B used the hops the inevitable happened and a complete brew, some 12,000 gallons, was ruined. The court decided that the statement by W was a term of the contract, even though not mentioned in the sale note. The statement that the hops were free of sulphur was known by both parties to be of considerable importance to B, and without that promise the contract would not have been made at all.

See also the following –

The SS Ardennes (1950) A grower shipped a quantity of oranges from Spain to the UK. The owners promised that the ship would sail direct to London. As the cargo was both highly perishable and intended for the seasonal market in the UK, this promise was obviously important. In fact, the ship sailed by way of Antwerp, arriving in London some weeks late, so that the grower lost a favourable market for his produce. The grower sued the owners for damages. The owners tried to rely on the standard contract terms in the bill of lading, which said that the ship could sail by any route and directly or indirectly. The court dismissed this argument and held that the oral promise made by the owners was a part of the contract as well as the terms of the bill, and took priority over the terms of the bill.

Importance and expert knowledge –

Dick Bentley Productions v. *Harold Smith Motors* (1965) DBP was the management company for the radio comedian Dick Bentley. He wanted to use a Bentley car. DBP approached ASM, a dealer in prestige second-hand cars, for a 'well-vetted' second-hand Bentley. ASM stated that they had an excellent car in stock which had in fact done only 20,000 miles since having a complete replacement engine and gearbox fitted. This statement was completely incorrect; the car had actually done well in excess of 100,000 miles since having a replacement engine and gearbox fitted, and was not in good mechanical condition at all. DBP bought the car. No mention was made of the statement made by ASM in the actual sale contract, but when a dispute later arose about the condition of the car the court held that the statement was a term of the contract. It was quite clear that if the statement had not been made then DBP would not have bought the car. In addition, ASM, as an expert in second-hand Bentleys, should have known, or could easily have checked, whether the statement was correct or not.

Esso Petroleum v. *Mardon* (1976) Discussing a lease of a petrol filling station, Esso told M that Esso estimated that within three years the throughput for the station should reach 200,000 gallons a year. M therefore agreed to take the lease. M was competent, but despite everything that he could do, the maximum throughput he managed to achieve within the three years was only some 70,000 gallons a year. As a result his business failed and the lease was terminated. In the resulting dispute M claimed that the statement by Esso of estimated annual throughput was a term of the lease. The court held that the statement was not a guarantee by Esso that M would be able to achieve a throughput of 200,000 gallons. But it was a statement made by a company who had special knowledge and skill. (In fact, 'estimated annual throughput' was how Esso benchmarked all its stations throughout the UK.) Therefore, although there was no guarantee that M could achieve throughput of 200,000 gallons, it was a statement that M should have been able to achieve something in the order of 200,000 gallons. In no sense could 70,000 gallons be described as in the order of 200,000 gallons. Esso was therefore held liable to compensate M for breach of a term of the contract.

Contrast these cases with the following cases involving the sales of second-hand vehicles:

Oscar Chess v. *Williams* (1957) W traded in his car in part-exchange. When asked by OC, the car dealer, for the age of his car he said that it was a 1948 model. The car was valued for trade-in purposes on that basis. But it was not a 1948 model, and had been manufactured in 1939. (Due to the Second World War there were no changes in car design between 1939 and 1948, and most cars were mothballed for the duration of the war.) W had relied on the date shown in the car's registration book, but this had been fraudulently altered by the car's previous owner. Clearly the age and trade-in value of the car was of serious commercial importance to OC; however, the court held that the statement made by W had not become a term of the contract. W was not expert, whereas OC was expert and also had much better ways of verifying the car's age than W.

Routledge v. *McKay* (1954) This case was not dissimilar to *Oscar Chess* v. *Williams*. M discussed the possible sale of a motorcycle with R, and, again relying on the logbook, gave the year of registration as 1942, when in fact it was 1930. A week later they met again and did the deal. The court held that the statement was not a term of the contract. All that M did, in fact, was to pass on to R the information in the logbook, and in any event the week's delay meant that R was not especially influenced by the statement.

So the principles are –

The further removed in time the statement is from the offer/contract, the less likely it is that it will be a term of the contract. In a private contract, as in *Routledge* v. *McKay*, a week was held to be too long. Of course, in a commercial situation, where discussions might go on for months, the time could be rather longer.

Where the person making a statement is not a professional or expert, or does not have special knowledge, and does not try to mislead, then the statement will not become a term of the contract, even though it influences the other party. But if he tries to mislead and succeeds in misleading the other party, the statement may become a term of the contract, or may be treated as a misrepresentation.

Where the person to whom a statement is made is not a professional or expert, he will not be expected to check the truth or correctness of any statement.

Where the person making a statement is a professional or expert or has special knowledge then he will be expected to know that any important statement will influence the other party to a significant extent in deciding whether or not to enter into the contract. Therefore, it may become a term of the contract, or it may be treated as a representation.

Where the person to whom a statement is made is a professional or expert, he will be expected to check the truth or correctness of any statement so far as it is reasonable for him to do so.

Where the party making the statement has special knowledge, so that the other party is not able to check the correctness of the statement, then the statement will probably become a term of the contract or may be treated as a misrepresentation.

The law treats a commercial person differently to a private person. In both *Bannerman* v. *White* and *Oscar Chess* v. *Williams* an incorrect statement was made in an honest belief in its truth. In both cases the statement was not true, because of the fraud of a third party. In *Bannerman*'s case a grower had lied, and in *Oscar Chess* a previous owner had falsified the registration book. In both cases the statement was significant, and was recognised to be so. In *Bannerman*'s case it clinched the sale, and in the *Oscar Chess* case it significantly increased the trade-in value of the car. However, where the statement was made by a commercial person/organisation, the court required the statement to have contractual status – effectively requiring the other party, White, to take the risk, whereas in the case of a statement by a private person, Williams, the reverse was the case.

The collateral contract option

Finally, even if the court does not feel able to find that an oral promise is to be included in an otherwise complete written contract, it may still be prepared to find a way round the difficulty where it feels that the circumstances justify it.

> *City and Westminster Properties (1934) Ltd* v. *Mudd* (1959) M was a desirable tenant. He was a highly reputable locksmith, who leased a shop and workroom in the City from CWP. He often had to work unusual hours, because he regularly helped the police to deal with burglaries and so on. At the rear of the shop was a small room which he used as a temporary bedroom if he had to work very late and so was unable to return to his home. In 1947 the lease became due for renewal. CWP inserted a new standard clause in the lease that the premises might only be used for trading, and must not be used for any non-trading purpose, which would of course prevent M staying overnight. (The purpose of this clause was to ensure that no tenant could in effect become resident at the premises, which would have made it much harder for CWP to terminate a lease.) M, quite properly, informed the land agent representing CWP that as his business often required him to work late, it would be impossible for him to renew the lease. The agent therefore assured M that if he accepted the lease as written CWP would raise no objection if he continued to sleep on the premises from time to time. (It was in fact doubtful whether the agent told CWP of this.) M therefore signed the lease 'as written'. Later CWP brought an action against M for forfeiture of the lease on the basis that he was in breach. M objected and the court accepted his objection.

The reason given by the court for its decision was not that the oral promise by the agent had become a term of the lease. Instead the court held that on the facts CWP had entered into two contracts with M. The lease was the first, and in addition a second (oral, collateral) contract had been made agreeing that, in return for M renewing the lease (which saved CWP the time and expense of finding a new tenant), the terms of the lease would not be enforced should M continue to sleep at the premises on an occasional basis.

6.2 Misrepresentation

The cases already mentioned in this chapter are concerned with misrepresentations that have resulted in claims for damages. Most predate the Misrepresentation Act of 1967, but their importance is that they show how the law deals with the problem.

Misrepresentation in general covers various remedies for situations in which someone has not told the truth during the discussions leading to a contract. As a result it is closely linked to other areas of law: the criminal law relating to fraud, tort law relating to negligence and deceit, the law of mistake, and so on. It is a difficult area of law to deal with, because it involves so many areas of uncertainty: knowledge versus suspicion; evidence versus knowledge; loss versus proof of loss; and so on. It is also rather complicated.

It developed as a confused mixture of equity and common law, bedevilled by the inability of judges to create a coherent set of principles. The result was that the remedies available to the innocent victim of a misrepresentation were both limited in extent and uncertain. This was largely remedied by the Misrepresentation Act, which established the basis of the current law.

There are, to put it simply, four questions that matter –

- What is an *actionable* misrepresentation (one for which the law will give a remedy)?
- When can the innocent victim claim compensation/damages for an actionable misrepresentation and on what basis?
- When can the contract affected by that misrepresentation be cancelled/rescinded?
- Can a contract exempt either of the parties from liability in respect of a misrepresentation?

What is an actionable misrepresentation?

A 'representation' is any statement made by one party to the other before contract with the aim of persuading that other party to enter into the contract.

A 'misrepresentation' is a representation that is not true/correct.

A misrepresentation may be –

Fraudulent – made without any honest belief in its truth. This was and is the classic definition. (It actually comes from *Derry* v. *Peek* (*1889*) a case not in contract but relating to the issue of a share prospectus by a railway company.) It may be made either deliberately (knowing it to be untrue) or recklessly (knowing that it may be untrue and not caring whether it is true or not). For an example of deliberate misrepresentation see the case of *Yam Seng Pte Ltd* v. *International Trade Corporation* (2013) above.

Negligent – made carelessly, without taking the proper steps to establish whether it is actually true or without believing it to be true. See the case of *Esso Petroleum* v. *Mardon* (1976) above (where Esso had failed to take any proper steps).

See also the case of *Howard Marine* v. *Ogden* (1978). This case concerned a contract to hire two barges, and in answer to a question the owner overstated the carrying capacity by a considerable margin from (faulty) memory, when he could easily have checked what the correct figure was. The court held that this amounted to a negligent misrepresentation, or rather a breach of an obligation not to give information without reasonable grounds for believing it to be correct.

Innocent – made with an honest belief in its truth. See the case of *Oscar Chess* v. *Williams* (1957) above.

To be actionable a misrepresentation must comply with the following –

It must be a statement of fact. A statement of opinion will not normally be enough to qualify, unless the maker is in possession of knowledge that makes that opinion impossible to justify. This is so even if the opinion is wildly incorrect; see the case of *Bissett* v. *Wilkinson* (1927) in which a farmer selling land produced an opinion that the land could pasture almost double the true figure.

A statement of opinion or intention will not normally qualify. But a statement of opinion or intention may amount to a misrepresentation if it amounts to a statement of fact or if it misstates the maker's true position; see *Edgington* v. *Fitzmaurice* (1885), for example, where a company issued a share prospectus apparently to invest in an expansion of its business, but in reality simply to raise the cash to pay off a major creditor. (And of course there will always be a problem when someone deliberately seeks to avoid making a representation, by stating it as 'merely his opinion'.)

It must be relevant to the contract that is being contemplated, and must be made during or as a part of the discussions leading to the contract.

It must be made with the intention of inducing the victim to enter into the contract. If a misrepresentation is made but does not influence the party to whom it is made, for example, because he does not believe it, then it will not be actionable.

In *JEB Fasteners Ltd* v. *Marks Bloom & Co* (1983) a takeover buyer relied on accounts that were wrong and had been prepared negligently. He later claimed damages. His claim failed because it became clear that his real objective was to take over some of the other company's people, so that any misrepresentation in the accounts did not influence his decision.

It must have the effect of inducing the victim to make the contract. It does not make any difference if the victim is also persuaded to make that contract by other factors as well as the misrepresentation.

There is no requirement for the victim to check for himself whether the (mis)representation is true. He is entitled to take it at its face value, especially if it is difficult for him to check or the maker is expert or professional. But if it is normal practice to conduct his own investigation, or his professional expertise leads him to check for himself, and he does so, then the inducement by the misrepresentation will cease –

Attwood v. *Small* (1838) A purchased a mine from S after S had given him false information about the quantity of ore remaining. However, A had also had an independent survey carried out, which had itself produced false information. The court refused to rescind the contract, since A had been influenced by his own survey, not the information provided by S.

It must be untrue at the time when the victim is induced to make the contract. If a representation that was correct at the time that it was made later becomes untrue, then it will become a misrepresentation unless corrected.

It must be made by one of the parties to the contract, or by an agent acting on his behalf. Statements by third parties, however persuasive they may be, are not actionable.

It will normally be made by words, but it is possible to give a misrepresentation by actions, such as by wearing army or naval uniform for instance.

Silence, the failure or refusal to give information, will not normally be a representation. However, the deliberate concealment of something such as a defect in goods will be a misrepresentation. Also, partial disclosure may amount to a misrepresentation –

> *The Spice Girls* v. *Aprilia WS BV* (2000) AWS wished to run an advertising campaign for its motor scooters. This would involve a considerable financial outlay, but the campaign would then run for several months. AWS settled upon the Spice Girls to front the campaign, for a considerable fee. At the time the Spice Girls knew that one of the members of the group was about to leave. The group did not reveal this and filmed the commercials. Geri Halliwell then left the group, which immediately reduced the commercials' value. The court held that the Spice Girls had made a misrepresentation that they intended to stay together for the period of the campaign.

Finally, where the contract is one of *uberrimae fidei* ('the utmost good faith'), such as an insurance policy, where there is a duty to disclose, failure to disclose will amount to a misrepresentation.

When can the innocent party claim compensation/damages for an actionable misrepresentation and on what basis? When can the contract affected by that misrepresentation be cancelled/rescinded?

The Misrepresentation Act extended and simplified the remedies available to the victim of a misrepresentation both to claim damages for his loss and also to have the contract rescinded.

The position is –

The victim is always entitled to recover damages for loss suffered as a result of fraudulent misrepresentation and of negligent misrepresentation. The victim must be able to produce evidence that –

- in the case of fraudulent misrepresentation, the representation was untrue, and that the maker of the representation either knew or suspected that the representation might be untrue, and

- in the case of negligent misrepresentation, the representation was untrue.

If a representation is untrue but there is no evidence of fraud, then the maker of the misrepresentation will be liable for negligent misrepresentation unless he can prove that he had an honest belief in the truth of the representation and also had reasonable grounds for his belief and that therefore the misrepresentation was innocent. This is a high standard of proof.

The victim is only entitled to recover damages for loss suffered as a result of innocent misrepresentation where it leads to a contract and has become a term of the contract, and the court is not prepared to allow rescission of the contract, but awards damages instead.

The level of damages for misrepresentation

The basis of any claim for damages is that usual in a claim in tort rather than the basis normal for claims under the law of contract. The basis for damages in contract is whatever sum will put the claimant into the position that he would have been in if the contract had been properly performed. For misrepresentation damages are whatever sum will put the claimant back into the position that he would have been in if he had not been induced to enter into the contract at all (often a much higher amount).

The new principles were originally set out in a case occurring soon after the Misrepresentation Act, *Doyle* v. *Olby (Ironmongers) Ltd* (1969), which was concerned with the proper level of damages payable in respect of fraudulent misrepresentation concerning the value of a retail ironmongers shop, and have since been confirmed by the House of Lords in *Smith New Court Securities* v. *Scrimgeour Vickers* (1996).

When can the contract affected by a misrepresentation be rescinded?

Rescission is not termination. Termination stops the contract at that moment. Every contract can be terminated.

But not every contract can be rescinded. Rescission means reversing the contract, so that it is as if it was not made at all. So everything paid has to be paid back, and everything delivered has to be handed back, as if the contract had never existed.

It is not possible where any third party has obtained rights as a result of the performance of the contract which would be prejudiced by rescission. It ceases to be available if it is not possible to restore the parties to their original position, say because work has been done which cannot be undone or goods have been supplied and then re-sold. It will also cease to be available if too much time has passed or the victim has affirmed the contract after learning of the true facts.

The current position now is that, subject to the above, rescission as well as damages is available as of right to the victim of a fraudulent misrepresentation.

Where the representation is merely negligent or innocent, rescission will only be available where the court is prepared to grant it. But if the court refuses to grant rescission it will award damages instead.

Exemption clauses and misrepresentation

Any contract clause that seeks to restrict liability for a misrepresentation shall be ineffective except to the extent that it is 'reasonable' within the definition of the Unfair Contract Terms Act 1977, section 8; see Chapter 9 below.

Note: The law is that where a commercial contract is made in writing, the written contract will be presumed to contain all the terms of that contract, unless there are very good reasons for allowing other terms to be included. Therefore the usual position in the commercial world is that the offer and acceptance process, *especially in standard situations* – sale/procurement, hiring/employment, and so on – is put into a procedure. The aim is to ensure that the company can control more or less exactly what goes into the offer. In addition, that offer will be put into writing, so that the contents of the offer can be identified with precision. The overall aim is to eliminate any uncertainty as to what the terms of any contract are.

7

Words in Contracts Part 2 – Post-Contract

7.1 Introduction

We need to be sure that we know what the contract actually says.

In the fascinating (to academic lawyers) case of *Smith* v. *Hughes* (1871), which was about the law of mistake, the decision depended upon whether an oral contract made between two farmers in a busy and noisy cattle-market was for the sale of 'good oats', or 'good old oats'. (Then, if the word 'old' had been said, was it part of the description of the goods, a condition of the contract, or a warranty – see Chapter 8 below.) The point was that both old and new oats make perfectly good porridge, but new oats cannot be fed to horses because they cause colic.

But once we know what the words of the contract actually are, they still have to be interpreted to decide the correct meaning of the contract.

The basic principles of interpretation are straightforward, although their application can be fraught with difficulty. They are –

Where a contract is made orally, or partly orally and partly in writing, the court will decide what the words of the contract are, based on the evidence given by the parties.

But where a contract is in writing, *and the written document or documents appear to set out all the terms necessary for a contract*, the presumption is that the written documents will be the whole of the contract. Oral statements will not be considered unless very clear evidence can be produced that the written documents were not intended to set out the whole of the contract. This evidence will generally need to be of the 'only reason why I was prepared to enter into the contract at all' variety. For examples, see the cases of *Bannerman* v. *White* and *The SS Ardennes* in Chapter 6 above.

Using Commercial Contracts: a practical guide for engineers and project managers, First Edition. David Wright.
© 2016 John Wiley & Sons, Ltd. Published 2016 by John Wiley & Sons, Ltd.

In the case of *Chartbrook Ltd* v. *Persimmon Homes Ltd* (2009) it was claimed that there was a mistake in a written contract and that it should be corrected in line with the pre-contract documentation used during negotiations. The court held that where there was clear evidence of 'a mistake on the face of the document', and of what correction ought to be made in the opinion of any reasonable person with full background knowledge of the circumstances, then the court was entitled to correct the mistake. In a situation where the parties could be shown to have a continuing common intention it would be proper to refer to evidence of previous communications between the parties that helped to show what they had really intended to say. But the usual rule was that previous negotiations should be excluded (because people can change their minds between negotiating and agreeing).

The only usual exceptions to this are where a contract is part of a series of contracts, or is directly related to other agreements, when they may be referred to.

The law presumes that the commercial organisation will have read the words that make up the contract, and will also have the commercial, technical and contractual knowledge to understand what those words mean. The law therefore presumes that the commercial organisation has agreed to the precise words that are written into the contract and accepts the obligations and liabilities that go with those words. This is actually two separate but related rules.

First, the commercial organisation is deemed to have read the words of its contracts, even if that is not the case in fact; see *L'Estrange* v. *Graucob* (1934). For the details see Chapter 3 above.

Second, the commercial organisation is also deemed to understand what the words of the contract mean when given their correct interpretation. Looked at in reverse this means that the commercial organisation must accept the risk that it might make a mistake when drafting the terms of its contracts, or might misinterpret the terms proposed to it by another party.

These two rules must be kept separate. The first applies to everyone, commercial organisation and private person. (It is subject of course to the rules on fraud and misrepresentation and so on, which are rather more protective towards the private individual than to the commercial organisation.)

The second rule applies differently to the commercial organisation and to the non-commercial consumer. The commercial organisation is required to accept virtually all the risks inherent in its contracts (subject only to a very small number of exceptions; see below). The amount of protection given to the consumer and employee by legislation is much higher.

The words used in a contract mean what a reasonable person knowing the factual background to the contract would think that they meant. Again, the different parts of this rule need to be considered as separate statements.

The reasonable person is not the average person. He is someone able to understand and evaluate the wording of the contract and the nature of any dispute that may have arisen, and also the evidence in that dispute, if necessary. He must also have enough technical expertise and knowledge of the commercial/business area or industry that forms the background to the agreement to be able to understand the context of the agreement, the significance of its terms, and the conventions, custom and practices that apply within that area or industry.

It goes without saying that the reasonable person must also be able to reach an impartial decision.

The factual background to the contract will influence the interpretation of the words. (For instance, in the cases of *Bannerman* v. *White* and *The SS Ardennes* [referred to in Chapter 4 above] the factual background would have included the fact that sulphur destroys beer and that the London market for oranges was highly seasonal and therefore price-sensitive.) Interpretation is a three- or four-stage process; see the case of *Investors Compensation Scheme Ltd* v. *West Bromwich Building Society* (1998) which describes the stages as follows –

1. Identifying all the background knowledge reasonably available to both or all the parties at the time that the contract was entered into. This includes anything that may affect the way in which the contract needs to be interpreted. This is sometimes called the 'factual matrix'. The factual matrix is primarily facts but could also include anything else that might affect interpretation. For instance, in *Crema* v. *Cenkos Securities plc* (2011) it was held that normal market practice should be included. But it does not include previous negotiations between the parties, or statements they have made to each other which have not become part of the contract. Nor does it include facts that are known to one party but not the other.
2. Deciding what is the actual meaning of the words of the contract. In doing so the words of the contract must be given their 'natural and ordinary meaning'; see below. And the basic principle is that the parties are deemed to have intended to use the words they did actually use.
3. Deciding what the parties should reasonably have understood those words to mean in the context of the factual matrix.
4. Then, if necessary, applying this interpretation to any actual or potential dispute that has arisen.

Remember that contracts can be very different. Some contracts contain very little information, and the only way to give them a commonsense interpretation is to look at the background. For instance, a contract with a bank for an overdraft may say nothing about why, or for how long, the borrower wants the money or why the bank has decided that it is a good risk. Borrowing to buy a car is very different from borrowing to set up a business. In such a case it is essential to look at the background.

Complex commercial contracts, on the other hand, often already contain within them much of the 'factual matrix'. In addition to information about what equipment or services are to be supplied, they will include information about how they are to be paid for, why they are needed, where and how they are to be used, and so on. The more information there is in the agreement the more important Stage 2 above becomes, and the less important Stages 1 and 3.

The aim is then to give the contract 'a common sense interpretation so as to give effect to the commercial purpose of the parties to the contract'. This is achieved by a careful analysis of the actual language contained in the contract.

7.2 The rules of contract analysis

English is a wonderful commercial language. It has a well-developed technical and commercial vocabulary. In addition, it is probably the most flexible of all modern languages. It is flexible because it contains a significant proportion of words with two or more quite distinct meanings, so that a comparatively small vocabulary can be used to express a large number of ideas. It also gives one many different ways of saying something. There are up to 20 different ways of saying 'I must go', though none of them will mean exactly the same as 'I must go'.

In commercial discussions that is a major advantage. However, in a contract the flexibility of English can become a serious disadvantage. It is all too possible for the words used in an agreement to have a meaning that is different to what was intended, so that drafting must be precise. Almost everyone who has been involved in a serious dispute concerning any complex contract/agreement will have found themselves faced with this type of problem.

In addition, the flexibility of the English language can create ambiguity. 'PER-FECT POLISH' can mean four different things depending on how we pronounce it. Consider a notice in a local newspaper in spring. 'Put your clocks forward one hour before you go to bed on Saturday night'. Now compare that with 'Before you go to bed on Saturday night put your clocks forward one hour'. The second statement has one meaning but the first has two. Changing the order of the words can create or eliminate a problem. The reason for the ambiguity in this example is that the phrase 'one hour' can either be attached to 'put your clocks forward' (by how much) or 'before you go to bed' (when). (Of course, one of these interpretations looks perverse as we all know what the newspaper meant to say. That is precisely the problem. Because we know what we mean to say we can sometimes fail to see exactly what we have said.)

7.3 An exercise in precise interpretation

The commercial contract is presumed to be a precise statement of the terms agreed between the parties. The problem with this statement, however, was well described by Lord Justice Diplock in *Hely-Hutchinson* v. *Brayhead* in 1968:

> Everyone outside a court of law recognises that words are imprecise instruments for communicating the thoughts of one man to another. The same words may be understood by one man in a different meaning from that in which they are understood by another and both meanings may be different from that which the author of the words intended to convey; but the notion that the same words should bear different meanings to different men and the more than one meaning should be 'right' conflicts with the whole training of a lawyer. Words are the tools of his trade. He uses them to define legal rights and duties. They do not achieve that purpose unless there can be attributed to them a single meaning as a 'right' meaning.

Therefore, if any contract ever has to be interpreted by 'the law', it will be given a precise meaning. This may be different to what the parties, or one of them, may have intended or believed.

The words of the agreement will also be given their correct grammatical meaning, unless this would produce a completely absurd result.

An example of this occurred in the case of *Matthew Hall Ortech Ltd* v. *Tarmac Roadstone Ltd* (1998). The agreement included a specification, written by TR. This stated that MH should do A and B in accordance with a standard specification. The court ruled that this meant that MH was required to carry out B in accordance with the standard specification. He also had to carry out A, but A did not have to comply. However, the judge then said that if the contract had said that MH should do **both A and B** in accordance with the standard specification … then his decision would have been different. Of course this decision may look difficult, in the sense that a strict grammatical interpretation has been allowed to prevail over the normal commercial intention, but the court had little choice but to give the decision that it did.

7.4 Some general rules

The words of the contract must, then, be given their natural and ordinary meaning. The natural and ordinary meaning will be that used among the general population, or among people involved in the commercial or technical field to which the agreement relates. It will, then, be the appropriate definition given in an authoritative dictionary.

If the dictionary gives more than one definition then a decision has to be made as to which of those definitions is to apply. (One of the clearest examples of a word that has several different meanings is 'tack', which has over a dozen totally different but equally natural meanings.) The court will decide this on the basis of 'normal English usage' among the relevant group. In the case of the commercial contract the group will be professionals, who give words their normal and correct (i.e. dictionary) meaning. Where the consumer is involved, normal and correct meanings will include colloquial meanings. If the court accepts that the proper meaning should be that used among a specialist group, the court will give the words their specialist, or jargon, meaning. (An example of this might be the word 'bit', which has ordinary meanings such as 'piece', 'part' or 'small quantity', and colloquial meanings such as 'a bit of a problem', but in the horse world means a part of a horse's harness, or in the world of oil exploration can mean the business end of a drill, and in the computer world an eighth of a byte. And in the American dialect, 'bit' can also mean a small coin.)

The meaning of words is a matter of fact (what the dictionary says) but the construction of a contract is a matter of law. The court will impose its interpretation upon the parties to the contract, whether they agree or not.

In addition, the court will generally adopt an objective approach. It will consider not what the actual intention of a party was, but what would have been the intention of a reasonable person in his position. This is, after all, what the other party has a right to expect, but means that the court's decision is independent of the wishes of the parties.

Preliminary contract drafts and negotiations may not in general be referred to when interpreting a contract. But where the parties have entered into a series of contracts, a previous contract may be referred to, if it will assist.

Often commercial contracts are based upon standard sets of conditions of contract. Standard conditions are usually of two very different types. First, they may be 'standard' or 'in-house' conditions, such as standard sets of conditions of sale or purchase, and so on. Second, they may be 'model' conditions of contract, which are generally accepted as a fair basis for contracts within a particular industry or area of business.

Where the court has to deal with model conditions of contract, it will be reluctant to disturb the general understanding and interpretation of those conditions within the industry, provided that the conditions are properly written. On this see the case of *Matthew Hall Ortech Ltd* v. *Tarmac Roadstone Ltd* above, relating to the process industry conditions published by the Institution of Chemical Engineers. This reflects the general approach of the law – that when the professionals have entered into a contract in the normal way, the law should not seek to disturb generally accepted commercial practice.

But where the court has to deal with terms which have been imposed by one party upon another, it adopts a different approach, often known by the Latin tag *contra proferentem*, that is, interpretation 'against (the interests of) the party that put the clause into the contract' and is usually trying to rely on it to avoid liability or gain an unfair advantage. *Contra proferentem* interpretation implies strict interpretation of the words. Only if the wording can withstand close scrutiny will it prevail.

This approach will generally be adopted by the court when considering in-house conditions of contract and any modifications to model conditions of contract that would have the effect of negating or severely disturbing the normal commercial position or interpretation of those model conditions, unless it can be shown that both parties freely agreed to those conditions.

Finally, where an agreement is made up of a standard or model form of contract to which the parties have added special conditions, greater weight will be given to the special conditions, and in case of conflict between the general conditions and the special conditions, the latter will prevail.

Words will always be given the meaning that they had at the time that the contract was signed. Therefore, if meanings change during the life of the agreement, say because of a change imposed by statute, the meaning of the contract will remain the same.

Words and clauses will always be read in context, not in isolation. The agreement must always be considered as a whole. As a corollary to this, all parts of the agreement must be given effect where possible. No part of the agreement should be treated as inoperative or surplus, but see below.

When an agreement has to deal with lists there are two principles that should be borne in mind:

- where the list expressly includes a number of things without any further comment, the inference is that other things of the same general category which are not expressly listed were deliberately omitted; this principle is sometimes given the Latin tag *expressio unius exclusio alterius*, 'saying one thing excludes the others';
- if an agreement lists a number of things of the same general category, but then does allow other things to be included (e.g. as is often the case in a *force majeure* clause), then other things will also be included but only if they are similar to those things already listed; this is sometimes given another tag, *eiusdem generis*, 'of the same kind'.

No contract will be construed, so far as possible, in a way that will permit one party to it to take advantage of his own wrongdoing.

Finally, we come to the problem of ambiguity. See above for two examples of ambiguity. In the normal commercial contract ambiguity is rare, but when it occurs it can be devastating in its effect.

A statement in a contract will be ambiguous when it has two or more meanings, each of which can apply without distorting the words.

If the ambiguity is latent, in other words, a hidden problem that only shows up when the parties try to carry out the terms of the agreement, then the court will try to find external evidence of what was intended by the parties to try to resolve the ambiguity. However, if the ambiguity cannot be resolved in this way, then the clause, or even the entire agreement, may become invalid, because its terms have been found to be uncertain.

A patent ambiguity exists when the contract, or any part of it, has two or more meanings right from the start. Where there is patent ambiguity in any part of the agreement the court can refer to subsidiary (lower-ranking) documents in the agreement, if they will help to resolve the ambiguity, or the court can take external evidence. But if this does not resolve the ambiguity then the clause or contract will be uncertain, and so invalid.

As a general principle, a contract, or a clause in a contract, will be uncertain if it is impossible to decide what it really means. If a clause or a contract is uncertain then that clause or contract is invalid. If it is invalid then it has no meaning. See the case of *Nicolene* v. *Simmonds* in Chapter 4, for example.

8

The Terms of the Contract

We have already seen in the last chapter that the law can bring additional terms into the contract from the previous dealings between the parties. This chapter looks at the other ways in which the law can impose extra terms on the contract, and the importance of the different types of term.

Additional terms are not imposed often. The aim is that the parties should make their own contracts, not judges. The only frequent cases involve terms implied by statute – especially the terms implied under the Sale of Goods Acts and their related legislation. Implication by legislation is one of the standard methods by which the state seeks to try to ensure a reasonably level playing field or promote public policy in the world of contracts.

8.1 Express and implied terms

The contract will be made up of *express terms* and also, sometimes, *implied terms*. The express terms are those specifically agreed between the parties. The implied terms are everything else that may apply.

In a contract made by the acceptance of an offer, the express terms will simply be the terms of the offer accepted as a package. Even in a large complex written contract that has resulted from negotiations between the parties so that the individual express terms will have been discussed and agreed in in detail, the theory is still the same – that when the negotiation is complete there will be an acceptance stage.

But even the most precisely drafted contract may at some point depend upon an implied term even if it only uses words which, for instance, refer to 'normal practice within the industry'. What the parties are doing is simply agreeing to leave something undefined, and therefore implied, and relying upon being able to define it later if needed.

Using Commercial Contracts: a practical guide for engineers and project managers, First Edition. David Wright.
© 2016 John Wiley & Sons, Ltd. Published 2016 by John Wiley & Sons, Ltd.

Law allows other terms to become part of the contract in addition to those specifically included in the successful offer or agreed between the parties. Terms may be implied into a contract for a number of reasons.

They may be terms that –

- the parties did have in mind but did not bother to put into words because they were so obvious – a typical example might be the purchase of furniture in a shop 'to be delivered', when it would be implied that the goods should be delivered by the shop reasonably soon after purchase;
- the parties are presumed to have had in mind but which they did not put into words; for an example see *The Moorcock* (1889) below;
- make a particular contract work properly; for an example see *Shirlaw* v. *Southern Foundries Ltd* (1939) below;
- make a class of contracts work properly or fairly; for an example see *Liverpool City Council* v. *Irwin* (1977) below; or
- the law thinks ought to be implied for the sake of fairness or public policy, such as the terms implied under the Sale of Goods Act.

The 'standard textbook' definition is that terms may be implied into the contract by custom, by the court or by law, by statute, or where a 'course of dealing' has been established between the parties.

8.2 Implied terms

Terms implied by 'custom'

Terms may be implied into a contract by custom – the customary practice within a particular business, trade or type of transaction.

In particular, a number of trading markets operate with their own rules, especially where they have to contend with large volumes of contracts made within tight timescales and cannot allow more than the absolute minimum time to agree each deal. Lloyds, the Stock Exchange and various commodity markets are obvious examples of this, each having their own customary terms of payment, settlement dates and so on.

If a term is to be implied into a contract by custom, it must first be proved that the custom does exist. The custom must be capable of being defined with reasonable precision and will normally need to be proved by expert evidence to be accepted within the trade or market in question.

This is a fairly obvious statement, but one which raises a very important point. If a party wants to rely upon what it believes to be normal custom, but is not certain that the custom does exist or is recognised as standard, it is better to write the custom into the contract to avoid any possible doubt.

Where a custom can be shown to exist it will apply, unless it conflicts with an express provision of the contract.

The custom on termination of agricultural leases (from the distant past) –

Hutton v. *Warren* (1836). Agricultural leases for tenant farms were generally based upon a one-year tenancy, running from the early-spring quarter day, often, but not always, renewed. W gave H notice to quit when his lease expired, and insisted, as he was fully entitled to do, that H should continue to farm the land until the end of the lease, but he then refused to pay H for the costs of sowing the summer crops. There was no mention in the lease of any specific conditions relating to the arrangements for termination. H claimed that it was customary to pay farmers quitting a farm at the end of the lease a fair allowance for the labour and seeds used to sow the summer crops. The court found the custom proved and held that a term requiring W to compensate H should be implied into the lease.

Second, an example of proven custom versus contract.

Les Affreteurs Reunis v. *Leopold Walford (London) Ltd* (1919) LW was a shipbroker who acted on behalf of a company which wanted to negotiate an extension to its contract to lease a ship named *The Flores*. Shipbrokers were paid commission by the ship-owner, in this case LAR. After the new lease had been signed (in 1916) but before it could commence, the French government requisitioned *The Flores* for use in its war effort. The lease was therefore barred from coming into operation. The evidence given by LAR was that the custom was for shipbrokers' commission to be paid by the owner out of his profits from the lease, in other words, only after the lease had commenced, and the ship was earning its keep. But in this case the lease stated specifically that the commission should be payable to LW on signature. The court decided that the customary terms of payment had clearly been proved but that they must be over-ruled by the express terms of the lease. LW was therefore awarded his commission.

Terms implied by law

This category covers two, quite separate, situations. Both are concerned with cases in which judges have acted to correct obvious injustice.

Terms 'implied by the court' in particular cases
The label is used to describe those cases where a term is implied into a particular contract. They are rare, and occur only when the court feels that action is needed to correct a particular situation that has gone badly wrong.
   Two of the classic cases are –

The Moorcock (1889) concerned an implied term to provide 'business efficacy' in a contract. A ship, *The Moorcock*, was to unload cargo at a wharf on the Thames. The contract was between the ship and the owners of the wharf. When the tide went out ships unloading at the wharf would settle on a mud-bank beside the wharf. *The Moorcock* was several feet deeper in draft than ships that had used the wharf before and when the tide went out she suffered major structural damage from a ridge of rock within the mud-bank. No mention had been made at any time of the possible risk of damage to the ship as a result of grounding. The court held that both parties were well aware that at low tide the ship would ground on the mud-bank. However, the owners of the wharf used the wharf as a part of their business, and would be expected to check what lay underneath the mud-bank. Therefore, when they hired out the wharf it must be implied that they had taken reasonable care to ensure that the wharf was safe for use by the ship. No ship would use the wharf unless the owners would give an undertaking that it was safe to do so. There was therefore an implied term in the contract that the wharf was safe for use.

Shirlaw v. *Southern Foundries* (1939) S agreed with SF to act as its managing director for a period of ten years. After some years the principal shareholder sold his shares. The new owner changed SF's articles of association to require directors to hold shares in the company and then refused to sell any shares to S, effectively ending his role as a director and therefore barring him from being managing director. S claimed damages for termination, and was successful. The court held that terms were to be implied into the agreement that the company would not remove S's directorship, or change its articles so as to permit anyone else to remove his directorship.

In *The Moorcock* the court used what it called a test of 'business efficacy'; that *no one* would use the wharf without such a clause being a part of the contract. In *Shirlaw*'s case, 50 years later, the court suggested a different test, that of the 'officious bystander'. A term should only be implied into a contract if, when a bystander suggested that it needed to be written into the contract, *both parties* would immediately dismiss his comment as being too obvious for the term to need putting in.

The two tests are clearly different, and the academic world has discussed the difference at great length over the years. But the terms have the same underlying idea, that a term should only be implied into a contract where it is absolutely necessary to prevent a wholly unjust result, and where it cannot really be disputed.

Both are very stringent tests, and have regularly been used since that time in cases where implied terms have been suggested, with the usual result that the proposed term has been rejected.

There are some terms that are regularly implied:

- that neither party shall prevent the other party from performing its side of the agreement;
- that where performance of the agreement cannot take place without the co-operation of both parties, then co-operation shall be forthcoming; and
- that where a contract does not fix a time for the performance of any obligation it is to be performed within a reasonable time.

The following case is an example of the type of situation where the court has been prepared to imply a term into a contract, in this case applying the 'officious bystander' test –

> *Greene Wood & McClean LLP* v. *Templeton Insurance Ltd* (2009) TIL had provided insurance to miners claiming damages for silicosis/emphysema against any non-recovery of legal costs, through GWM, who was representing them. Was a term to be implied into the policy that it would also cover GWM as well as the miners who were its clients? The court decided that the term should be implied. It must have been obvious to TIL that GWM would be giving undertakings to its clients that their costs would be covered by the policy, and that these costs would include the work done by GWM on their behalf.

Terms 'implied by law' for reasons of public policy

This happens when the court has decided that for reasons of public policy a term should be implied into contracts in a particular category, such as buyer/seller, landlord/tenant or employer/employee. Usually such terms only apply if they are not excluded or over-ruled by an express term of the contract.

(These terms are often subsequently adopted by legislation, the best examples of this being the terms implied under the Sale of Goods Acts.)

These terms are more common, and have much wider application, although again the number of actual cases in which the courts have found that they exist is small. The classic example concerns council flats in Liverpool. It also demonstrates perfectly how the law sets out to solve the difficulties of reaching a fair decision in a new and general situation.

> *Liverpool City Council* v. *Irwin* (1977) In 1967 LCC leased a flat to Irwin on the ninth and tenth floors of a high-rise council block. Access to the flat was by two lifts and an enclosed stairway. The flat was also served by a refuse chute down which light rubbish could be dropped to ground level for disposal. There were several problems with the flat and the block as a whole. The lavatory was defective and flooded the flat with depressing regularity. The lights on the stairs and the lifts failed, and the chute became blocked. This was largely due to vandalism by other residents and third parties.
>
> The LCC agreement signed by I was not complete. It listed Irwin's obligations, but not those of LCC.
>
> After several years Irwin began to withhold rent in protest against the conditions, and eventually LCC terminated his lease and sought to repossess the flat. Irwin disputed repossession and counter-claimed for breach by LCC of their obligations.

The case was obviously important. It was one of the first cases in which the courts were given the opportunity to look at the obligations of a local authority as opposed to a private landlord. The case was therefore taken all the way to the House of Lords. The House of Lords looked in detail at the problem, examining the LCC's procedures and the documentation used for granting tenancies, visiting the actual tower block involved to see the problems at first hand and so on, before issuing their decision.

The court then decided that in relation to the defective lavatory the LCC was no different to any other landlord. It had an obligation under section 12 of the Housing Act of 1961 to provide proper sanitary facilities. It also had an obligation to use 'reasonable endeavours' to maintain the access routes in reasonable condition and refuse disposal chutes in good order. (The court then did find that in the circumstances the LCC had used reasonable endeavours to do this.) The court then ruled that these two terms should be implied into all LCC leases.

Terms implied by statute

Terms are implied into contracts by a range of different statutes of the English and Scottish parliaments, the Welsh assembly and so on, usually as a method of instituting public policy. Many types of contract are controlled in this way: employment contracts, consumer contracts, landlord and tenant contracts, contracts of sale and hire purchase, and credit sale contracts, to mention but a few. There are far too many to deal with them all in detail.

But we shall deal, in Chapter 9 below, with the most important implied terms for commercial contracts, those relating to sale and supply contracts.

8.3 Express terms

Express terms dealing with the effectiveness of the contract

The contract may contain terms dealing with the validity or effectiveness of the contract.

'Subject to contract' and 'condition precedent' terms

The parties will have made an agreement which contains a term that prevents it from being effective. This can be done in two different ways.

First, the agreement may be 'subject to contract'. This means that it is dependent upon a further step by the parties, such as the execution of a formal contract, before it can come into force.

Second, the contract may be subject to a condition precedent. In this case the contract will be in force, but its operation will be suspended until something necessary has happened or been finalised. One example that often occurs is in export contracts, when the two parties have entered into a contract but the purchaser now needs to finalise arrangements for the documentary credit or credit finance to enable him to pay the price under the contract. The contract will then contain a clause providing that it will only become effective when the purchaser has made his arrangements.

The reasoning is the same in both cases. Something needs to happen before the agreement will work. But the legal principles are very different. Of course, in most normal circumstances the distinction between the two should not cause any problems. Very often there will be no practical consequences between one approach and the other. Also, in commercial contracts the parties will be quite clear about what they will want to do and will be careful to use language that will achieve it, as in the example referred to above.

But this has become an area where the law has made difficulties for itself by trying to be too precise about whether a practical situation falls into one category or the other. Sometimes badly worded clauses or bad contract procedures also cause problems. Sometimes, however, legal logic has simply tied itself in knots. For example –

> *Bentworth Finance Ltd* v. *Lubert* (1968) Bentworth entered into a hire-purchase contract with L relating to a car. L was due to pay for the car in 24 monthly instalments. But the car was delivered to L by the dealer without a logbook, and despite all L's efforts no logbook was forthcoming. As he did not have the logbook he could not tax the car, and so could not use it. He therefore refused to pay the instalments. Bentworth reclaimed the car and sued L for payment. The court held that Bentworth was not entitled to sue L for any payments under the contract, which was quite clearly the correct decision.

However, the reason given by the court was that there was no enforceable contract at all between the parties until the logbook had been supplied. This creates the ridiculous situation that L had no right to the car for which he had contracted and which had been delivered to him simply because a third party, the dealer, was at fault. A much better decision, and one which we believe to state the law be correctly, is the case of *Trans Trust SPRL* v. *Danubian Trading Co Ltd* (1952). This case concerned a contract for the sale of steel products. It was a condition of the contract that the buyer should open a letter of credit for the price immediately after contract. He failed to do so. The court held this breach simply released the seller from any obligation to deliver the goods. There was an enforceable contract, but the lack of the proper existence of a channel of payment for the goods in breach of the terms of the contract was sufficient to allow the supplier to refuse to carry out his side of the contract.

Condition subsequent terms

A condition subsequent is a somewhat archaic description of a term in a contract that will permit one of the parties to terminate the contract in the event that a particular event occurs. The classic example is a case relating to the sale of a horse –

> *Head* v. *Tattersall* (1871) H bought an expensive horse from T to use to ride to hounds. T gave H a guarantee that the horse had hunted with the Bicester hounds,

in other words, that it was an experienced hunter, and that if this was not so that H could return it. The horse suffered an accidental injury, and when H discovered that the horse had, in fact, never been ridden to hounds before, he sought to return it to T and recover the price. The court held that H was entitled to do so. He was allowed to cancel the contract, recover the price and return the horse despite its injury.

8.4 Conditions, warranties and innominate terms

This is a very difficult, confusing and confused area. Unfortunately it is vital when writing and managing commercial agreements, especially those covering long-term relationships. The difficulty is caused partly by the way the law has developed, and partly by a complete split between the interests of the commercial organisation and of the judiciary.

The contract may include three different classes of terms –

- the 'condition' or major term;
- the 'warranty' or minor term; and
- the 'innominate' or 'intermediate' term, which comes somewhere in between the other two.

The only practical difference between these three types of term lies in the remedies that are available to the injured party in the event of breach. The law will always allow an injured party damages in the case of breach by the other party of any term of the contract. But the law will only allow the injured party to terminate the contract where there is a breach of a condition (however trivial that breach might be), or where there is a serious breach of an innominate term. Termination is not available for the breach of a warranty, even if that breach is serious, or for a less serious breach of an innominate term.

So there are several questions, of which the three most obvious are: First, when is a term of the contract a condition? Second, when is a term of the contract an innominate term? Third, when is a breach of an innominate term serious? The questions are simple. The answers are not. To understand what the law is now it does help to know how it has developed.

Development of the classification of contract terms

The modern system of classification of contract terms developed over a long period. To a large extent there was no real system up to the end of the 1800s. It was only in the original Sale of Goods Act in 1893 that a distinction was first drawn between condition and warranty. The distinction was useful but soon began to cause practical problems.

A good example of the early approach of the courts was given by two cases which came to court in the same year (1876) and involved very similar circumstances – *Bettini* v. *Gye* and *Poussard* v. *Spiers and Pond*.

Signor Bettini and Madame Poussard were both opera singers who were engaged to sing in England. Bettini was to give a series of concerts and recitals, Poussard to perform in an operetta at a London theatre. Bettini's contract required him to be available for rehearsals a week before his first performance, but due to a minor illness he was three days late. Poussard was to arrive in time for her first night, but due to a last-minute illness she arrived a week late and another principal soprano had to be engaged. The respective managements terminated both of their contracts for breach, and in each case the performer sued for wrongful termination.

In Bettini's case the court found that the termination was unjustified, describing the time clause in the contract as a warranty, but with the minimum of analysis. In Poussard's case the court decided that the time clause was a condition and that therefore the termination was justified.

Of course, these two cases raise difficulties. The decisions were made without any detailed discussion by the courts as to what might constitute a condition or a warranty in relation to time. They also look just a little bit too sensible, which raises the suspicion that perhaps what was happening was reverse logic. Were the courts looking at the seriousness of the breach, deciding what a reasonable result ought to be and then interpreting the words of the contracts in a way that would produce that result?

The division of contract clauses into conditions and warranties raised another issue. The inevitable implication of classifying any particular provision as a condition is that any breach of that provision, however minor, will entitle the injured party to terminate the contract should he wish to do so. Inevitably this can produce harsh decisions. In one case for example, *Arcos Ltd* v. *E A Ronaason & Son* (1933), a contract for the sale of barrel staves was terminated successfully on the basis that staves described as half-an-inch thick were actually marginally thicker, by less than an eighth of an inch, than the description stated in the contract. (Under the 1893 Sale of Goods Act, as in the current Act, in a sale of goods 'by description' it was a condition that the goods supplied must comply with the description.) The staves were perfectly usable. There is no way of knowing now why the buyer did not want to accept them. There may have been a good technical reason, or he may have simply wanted to escape the contract.

Another oft-quoted example is the following –

Moore v. *Landauer* (1921) This case concerned the sale of a large quantity of tinned fruit by an Australian supplier, M, to L, an importer in London. The contract stated that the goods, 2.5 lb tins of pears and peaches, were to be packed in cases each holding 30 tins. When the goods arrived in London it was discovered that rather more than half had been supplied not in 30-tin cases but in 24-tin cases. L rejected the goods and M sued. The Court of Appeal held that this was a sale by description, and that the packing requirements were a part of the description of the goods. Tins packed in 24-tin cases were of a different description than tins packed in 30-tin cases, even though the tins were precisely the same. Therefore L was within his rights to terminate the contract for breach.

This decision was later criticised as an example of how unfair the principle of the condition can be in practice. But what was under consideration here was the import trade. Within the trade it was well known by both suppliers and buyers that goods of this kind would be fed by the importer straight into the distribution chain. So the packing of goods could be of enormous importance for their resale. It was also normal practice for the importer to have sold the goods on to distributors long before they arrived within the UK. Therefore M should have known perfectly well that the packing of the goods was important to L. M had clearly breached this requirement and to a considerable extent. In the circumstances it was perfectly proper to treat the clause as a significant part of the description of the goods, and therefore as a condition.

Then a series of cases arose relating to second-hand cars during the 1950s and 1960s, as ownership increased following the Second World War. Typically, the buyer would be dissatisfied because the car was not what it was supposed to be. There would be no doubt that the vehicle did not meet its description or was defective, or that the defects were a breach of the contract, or that the buyer had a valid claim for damages. The point in the dispute was whether the buyer had the right to return the car to the dealer and get his money back. He could do this if the breach was of a condition. But if the breach was only breach of a warranty, he would have to accept a small sum in damages.

The inevitable result was that the courts delivered a series of judgments containing complex arguments as to whether particular words in a particular contract were to be classed as a condition or a warranty. What the courts were actually doing, but would not admit, was again applying reverse reasoning to the facts. If the court felt that the breach was sufficiently serious to merit giving the buyer the right to terminate the contract it would classify the term as a condition. If it felt that this was not so it would classify the term as a warranty. Not surprisingly, this led to a number of fair but legally somewhat dubious decisions.

The 'solution' to this problem was created in a case in the Court of Appeal relating to a time charter, an agreement to hire a ship, for a period of two years. The case is extremely important because it created, or perhaps rediscovered, the innominate term. It is also important because the court made a number of statements about the nature of the 'contract condition' which led to further clarification of the current position by the Court of Appeal and the House of Lords.

Hong Kong Fir Shipping Co Ltd v. *Kawasaki Kisen Kaisha Ltd* (1962) HKF was a company that owned a single ship. It entered into an agreement with KKK for the lease of the ship to carry goods to and from Japan, carrying components and materials to Japan and then complete motorcycles from Japan to KKK's export customers. Among many other provisions the contract stated that the ship should be fit for ordinary cargo service, and should also have a competent crew. Immediately after the hiring commenced the ship was damaged on leaving port through the negligence of the crew. Having being repaired, she loaded a cargo to be carried to Japan and then the steering gear broke down and required further repairs. She then sailed to Japan. It was then discovered that she needed another three months in dry dock for additional repairs to her boilers.

In effect the first five months of the charter had resulted in one single completed voyage of some two to three weeks' duration. At this point KKK terminated the contract for breach, even though it would still have had some 19 months left to run once the repairs were complete. KKK had clearly been advised that it was entitled to treat the terms of the contract concerning the efficiency of the crew and the state of the ship as conditions.

HKF now sued for damages for unfair termination of the contract. The court decided that these terms were neither conditions nor warranties, but that they were innominate terms which would justify termination of the contract only in the event of serious breach. If the breach was not serious then KKK would only be entitled to claim compensation for the cost of hiring other shipping but would have to allow HKF to carry out the rest of the contract hire. The court then decided that the breach that had occurred was not sufficiently serious to entitle KKK to terminate the contract. HKF was therefore awarded substantial damages for unfair termination of its charter.

There are a number of comments to be made about this decision.

In theory, the innominate term is a perfect solution to the problem of the rights of the injured party in every case of breach of contract. In the immediate aftermath of the *Hong Kong Fir* case, learned articles appeared in several legal journals prophesying that the innominate term would soon make not only the warranty, but also the condition, redundant. This did not happen.

It is quite clear that Kawasaki was in an almost impossible position. It had virtually lost the first five months of its charter. It had been forced to charter alternative shipping space at short notice and high cost and for a substantial period, and it had probably lost all confidence in the ability of Hong Kong Fir to carry out the rest of the charter properly. It had to make a decision. To allow the charter to continue was to risk another disaster. Therefore it took the decision to terminate, believing that it was entitled to do so. Four years later it was now being told by the Court of Appeal that it had been wrong, and as a result incurred heavy damages, without actually being in any way to blame.

It has to be said that the quality of the decision made by the trial judge and supported by the Court of Appeal on the facts that the actual breach was not serious enough to justify termination was questionable. The reason given by the court was that as the contract still had well over a year left to run at the time it would be wrong to strike it down since Kawasaki could produce no evidence to show that Hong Kong Fir could not have performed. (Perhaps what the trial judge should have done was to simply reverse the burden of proof and require Hong Kong Fir to prove reliability.)

Compare the decision in the following case –

Farnworth Finance Facilities v. *Attryde* (1970) In July A bought a high performance motorcycle from a dealer. The purchase was financed by FFF by hire purchase over 24 months. The motorcycle developed a series of defects that made it quite dangerous to ride. In November A repudiated the contract. The defects included a pannier falling off and causing a skid, the headlight failing while the motorcycle was being ridden at night, chronic instability, and finally the rear drive chain breaking. Luckily this last failure occurred at low speed so that A was not injured. In this case the court had no difficulty in deciding that the series of defects was quite enough to entitle A to terminate the hiring agreement after four months of poor performance.

The Hong Kong Fir *case and conditions*

In the *Hong Kong Fir* case the Court of Appeal defined a condition of contract as including only those terms of which every breach, however small, must inevitably deprive the injured party of substantially the whole benefit that he would expect to gain from the contract.

This definition would have limited contract conditions to only a tiny number of situations. It was almost immediately recognised as mistaken, and was challenged and discounted in a series of cases in the Court of Appeal and the House of Lords which discussed and clarified the position.

In the first of these, *The Mihalis Angelos* (1971), the dispute concerned a contract made on 25 May to load a cargo for export from the port of Haiphong in Vietnam. Under an 'expected readiness' clause in the contract, the ship-owner stated that she would be ready to load cargo at Haiphong around 1 July. In fact, this statement was wholly untrue. The earliest that the ship could in practice have arrived at Haiphong was 25 July. (But it was quite clear that if the ship had arrived at Haiphong a few days late, the breach would not have deprived the owner of the cargo of substantially the whole benefit of his contract.)

In fact, the cargo owner acted very reasonably and waited until 17 July before terminating the contract. The ship-owner sued for damages. The Court of Appeal decided that the readiness clause was clearly a condition of the contract and upheld the cargo owner's right to terminate. It was essential within the shipping industry for a cargo owner to know well in advance when the ship would arrive and where. After all, he might have to deliver several thousand tonnes of cargo to the port in time for it to be loaded. He must be able to terminate the contract if the ship did not arrive on time so that he could make alternative shipping arrangements.

Then in another case, *Reardon Smith Line Ltd* v. *Hansen-Tangen* (1976), a case concerning the description of a ship being built in Japan, the Court of Appeal discussed the nature of a condition. The facts were that the ship was properly built to the detailed specification stated in the contract. However, the contract stated that the ship was to be built in a particular yard with a particular reference number. In fact, it had a different reference number and was built in a different yard. The customer terminated. The court held that the reference number was merely intended to identify one ship from a number of others and therefore was not a contract condition. However, the court then went on to say that where any particular item of the description was a 'substantial ingredient of the identity of the thing that was being sold', then it should be treated as a condition of the contract.

Finally, we come to a case decided by the Court of Appeal and then confirmed by the House of Lords.

Bunge Corporation v. *Tradax Export SA* (1981) Under a contract for the sale of 5000 tonnes of soya flour to be delivered Free on Board (FOB) Gulf (of Mexico) port by 30 June, the buyer was to give at least 15 days' notice of when the ship (or ships) would arrive in the Gulf, so that the seller could nominate the port at which the goods would be loaded. The buyer did not give notice until 17 June, less than 15 days before the end of the delivery period. In theory it might have been just possible to load the goods within 13 days. On 20 June the seller terminated the contract and claimed damages. (The reason why he was actually happy to terminate was that the market price had fallen so that he could no longer sell at a profit.) The judge at first instance held that that the requirement to give notice was an innominate term but the Court of Appeal disagreed. The requirement was a condition of the contract. The seller needed to know when ships would be available so that he could carry out his side of the contract properly, and in the commercial contract the time promise is usually to be treated as a condition. The case was then appealed to the House of Lords. The House agreed that the *Hong Kong Fir* case was 'very valuable', in that it gave greater flexibility to the courts in deciding how to deal with any particular problem, but it decisively rejected the idea that there should be severe limits upon the ability of the parties to a contract to define terms as conditions where there were important commercial or technical reasons for having a right of termination. The parties to commercial contracts must be able to insist on certainty when they have clearly demonstrated in the contract that that is what they intend.

The effect of these decisions has been to widen the definition of clauses that are to be treated as contract conditions from only those terms of which any breach however small would deprive a party of substantially the whole benefit of the contract. The definition includes any term which is stated in the contract to be a significant or substantial element of what is being sold or supplied, or is necessary to provide the proper degree of commercial certainty in the business concerned, or that relates to any other major requirement of the contract.

What is a warranty?

A warranty is simply any term of the contract that is not treated by the parties as especially important in any way. Breach of the term will probably not have any substantial impact upon the overall purpose of the contract. The principal remedy for breach is damages.

What is a condition?

A number of different things will form conditions –

- first, the principal time obligation;
- second, all the important requirements of what is to be sold, supplied or done by either party;
- third, any other obligation described by any statute as an implied condition of the contract;
- finally, any other terms of the contract that are specifically described by the contract as conditions, and which would be agreed as being of significant technical or commercial importance.

What is difficult to decide is when clauses relating to the 'other terms' will be held by the courts to be conditions. The probable situation is that this will be the case where –

- the contract states very clearly indeed that a particular requirement is to be a condition of the contract; and/or
- the contract states clearly that the 'innocent' party shall have the right to terminate the contract in the event of any breach of that clause; and
- there is nothing in the rest of the contract that could be construed as con-tradicting these statements (see *Wickman*'s case below).

The principal remedies for breach, at the option of the injured party, are either termination or damages, or both.

What is an innominate/intermediate term?

Innominate or intermediate terms are all other terms of the contract (that are neither warranties nor conditions). The distinguishing feature of the innominate term is that although it is not described as a condition of the contract, it is possi-ble to see that a major breach could result in significant loss to the injured party. The principal remedy for breach is damages unless the breach is by nature seri-ous, in which case termination may be available in addition.

However, the problem is in deciding when a breach is serious enough to trigger termination as a remedy. It is clear, we believe, that if termination was challenged it could only be justified when the injured party could produce convincing evidence of a substantial loss of the benefit that was to accrue from the contract.

Trying too hard

The difficulty is that the parties may need to be careful in their drafting of the contract to achieve the desired result –

> *Schuler (L) AG* v. *Wickman Machine Tool Sales* (1972) SL manufactured presses used in the manufacture of bodywork panels for cars and commercial vehicles. It entered into an agreement with WMT to act as a sales agent for its products within the UK. Clause 7 in the agreement stated that it was a 'condition' of the agreement that WMT's representative should visit each of six major car producers in the UK each week to look for orders. (No other clause was described as a condition.) The agreement also included a termination clause which provided that either party could terminate if the other party was in significant breach of any of its obligations under the agreement and failed to remedy the breach within 60 days of being required in writing to do so. During the first six months WMT committed several material breaches of clause 7, but SL waived its right to terminate. During the next six months WMT committed further minor breaches, but for perfectly good reasons. SL, however, was becoming dissatisfied and finally invoked the termination clause after 17 months. WMT sued for damages for wrongful termination. The court held that SL was not entitled to terminate since there had been no significant breach of clause 7 for many months.
>
> SL then claimed that clause 7 was a condition and that therefore any breach, however minor, would justify termination. The court responded that the wording of the termination clause made the meaning of clause 7 ambiguous. By saying that only a significant breach of any obligation of WMT under the agreement, including clause 7, would justify SL in terminating the agreement, the termination clause had in fact changed clause 7 from a condition to an innominate term.

(In other words, Schuler tried too hard. By combining a 'condition' clause, clause 7, with a termination clause that referred only to significant breach, it managed to reduce a condition to an innominate term.)

Finally

It is always going to be uncertain what degree of breach of an innominate term would be accepted by the courts as sufficiently serious to entitle the injured party to terminate the contract.

This uncertainty will increase according to the duration of the contract. In an agreement covering a single transaction a serious breach of performance will be sufficient. In an agreement with a duration of, say, three months, a serious breach of two or three weeks would also presumably be sufficient.

However, in the *Hong Kong Fir* case disastrous performance extending over a period of almost five months was held not to be a sufficiently serious breach to justify termination of a 21-month contract. This raises a very serious question. How serious and of what duration must a breach of an innominate term be to justify the termination of a long-duration contract? In the modern business climate it is quite normal for companies to contract out the supply of equipment or services on a long-term basis, often over five-year, ten-year, or even longer, periods. It is probably true to say that to qualify for termination a breach of an agreement for so long a period might have to be of absolutely cataclysmic proportions or of enormous duration, or both.

The overall result is that the longer the duration of the contract, the greater the importance of clarifying the position as to which of the terms are classified as conditions rather than innominate terms or warranties, and of setting out clearly when and in what circumstances each party will be entitled to terminate for breach. Alternatively it might be appropriate to break up a long-term contract by creating options to terminate by the parties at regular intervals.

And, given the potential for damaging and costly legal disputes over any contested termination, it could well be advisable also to include a termination for convenience provision anyway, plus a clause specifically allowing for termination for any breach, thus allowing the parties to escape from the agreement without having to risk the possible consequences of a disputed termination.

9

The Basic Framework – Contracts of Sale

9.1 General

Many commercial contracts are regulated by law. Employment contracts, hire purchase contracts, credit sales and consumer contracts are controlled in the interests of protecting the individual. Intellectual property licences and agreements to set up distribution networks are regulated to control unfair competition. Contracts in the construction industry are regulated to control unfair payment practices and to reduce the impact of dispute.

But there is one contract for which legislation goes much further, and provides a comprehensive set of basic rules – the contract for the sale of goods. The rules are set out in the Sale of Goods Act and other related Acts. The rules are not just important for contracts of sale. They also affect, to a greater or lesser extent, the whole range of procurement/supply contracts, such as hire/hire purchase, the supply and installation of equipment, 'work and materials' contracts, building/construction contracts and contracts for the supply of services.

The process began with the first Sale of Goods Act in 1893. This was then replaced by the current Sale of Goods Act in 1979. The principal related Acts are the Supply of Goods (Implied Terms) Act of 1973, the Supply of Goods and Services Act of 1982, and the Sale and Supply of Goods Act 1994.

9.2 Preliminary points

Contracts vary depending upon the circumstances, but, broadly, the simpler the contract the more the obligations of the parties will be those laid down by the Sale of Goods Act. As a contract becomes more complex, its specific terms take over and replace the basic Sale of Goods Act rules. The result is that most

Using Commercial Contracts: a practical guide for engineers and project managers, First Edition.
David Wright.
© 2016 John Wiley & Sons, Ltd. Published 2016 by John Wiley & Sons, Ltd.

of the cases in which the basic rules are explored are either consumer contracts or trading contracts dealing with the sale of commodities.

The purpose of the contract of sale is to transfer or 'vest' ownership of, or what the law may call the 'title to' or the 'property in', goods from the seller to the buyer in return for the payment of a price.

The seller will be either the owner of the goods, or someone else who has the authority to pass ownership to the buyer.

'Goods' is a generic term. It includes anything and everything moveable that can be delivered by the seller to the buyer. 'Delivery' in Sale of Goods Act language does not mean delivery to anywhere. It means making the goods available to the buyer at a particular point so that the buyer can collect them. It may also mean an act that enables the buyer to take control of the goods, for example, handing him the keys and logbook for a car, or the shipping doc-uments/bill of lading for goods in transit, so that he can either sell them on or collect them when they reach their destination.

The price may be decided in different ways. It may be a fixed amount stated in the contract, or an amount calculated by reference to the quantity/quality of goods supplied, or the amount of work done, or even simply a 'reasonable price', possibly to be fixed by an agreed procedure set out in the contract.

9.3 The contract of sale and similar transactions

The contract of sale can be distinguished from various other similar transac-tions. In the past there were several important technical and procedural differ-ences between them, such as requirements for certain types of contract to be evidenced in writing. These could make the proper classification of the trans-action important. They have now largely disappeared except for some minor differences (see below).

Gift

A gift transfers the ownership of the thing from the giver to the receiver. But the promise to give cannot be enforced, unless it was made by a deed.

Deed

A deed is a formal undertaking which can be enforced by the person in whose favour the deed has been made. The enforceability results from the formality of the undertaking. So a promise by deed to pay or transfer something is binding even without consideration.

Of course, a contract can be made as a deed. Government, national and local, often insists that supply contracts are done as deeds (or for a company made under seal or signed as a deed). There are two principal reasons for this. The greater formality of the procedure means that approval for the contract must be confirmed at board level, and the limitation period for litigation/arbitration is 12 years for a contract under seal rather than the 6 years for a contract that is merely oral or signed.

Bailment

Bailment is not a contract but a relationship of trust. It is created when the owner or possessor of a thing, the 'bailor', gives possession of that thing to another, the 'bailee', with specific instructions as to what the bailee must do with that thing.

Bailment is often created by a contract. Typical examples of contracts for bailment are contracts to transport or repair something.

Barter

Barter is a contract for the exchange of goods or services in return for each other, rather than goods for a price.

Exchange

A contract of exchange is concerned with exchanging goods for units of value. A typical example is the exchange of savings on a discount card or tokens for goods. They are regulated by the Consumer Credit Act and the legislation referring to consumer sales contracts in general.

Lease/hire

A contract for the lease or hire of something grants the lessee, in return for payment, the possession but not the ownership of that thing for the period of the lease with the right to use it for whatever purposes are allowed by the lease. When the period of the lease expires the thing is returned to the owner.

Hire purchase

A hire purchase contract is a contract to hire something coupled with an option for the hirer to buy it, usually at the end of the hiring period. Hire purchase contracts have been made the subject of separate legislation starting with the Hire Purchase Act 1964 and the Supply of Goods (Implied Terms) Act 1973, which create similar obligations to a consumer sale contract.

Contracts for services

Contracts for services do not transfer the ownership of any goods from one party to the other, except purely incidentally. They are governed mainly by the Supply of Goods and Services Act 1982.

Construction/building/installation contracts and contracts for 'work and materials'

Finally, there is a large group of contracts for building work, etc., or for the installation of equipment, which result in the transfer of ownership of the materials and equipment, not by being delivered as in a contract of sale, but by their being permanently fixed in place and so becoming part of the structure or the land. These contracts are governed by an amalgam of the Sale of Goods Act and the Sale and Supply of Goods and Services Act, which produces a position very similar to that of a contract of sale.

The test of whether anything has become a part of a building or land is whether it can be removed easily. If, for example, it can simply be unbolted or unscrewed so that it can be removed without causing any damage, then it is not part of the building or land. But if it is, for example, permanently welded or cemented in place, so that it can only be removed by damaging the structure or land, then it is part of that land/structure.

9.4 The sale of goods versus the supply of services

In general terms the underlying basis of the obligations of the seller and buyer of goods is that those obligations are strict. The failure by the seller without a lawful excuse to supply goods that comply with the contract will automatically be a breach of the contract. So will a failure by the buyer to pay for or to accept the goods. To try one's best but fail is still a breach.

In a contract for the supply of services the obligation of the supplier is still strict. But the level of that obligation may be rather lower. Contrast the absolute obligation under section 14 of the Sale of Goods Act (as amended) for goods to be 'of satisfactory quality' and 'fit for purpose' with section 13 of the Supply of Goods and Services Act 1982, 'In a contract for the supply of a service where the supplier is acting in the course of a business, there is an implied term that the supplier will carry out the service with reasonable care and skill'. In other words, the supplier will have a strict obligation to supply the service but his strict obligation in respect of the quality of that service or the results to be achieved or produced by that service is not absolute but merely relative to the level of the reasonable supplier. In other cases the supplier will not be liable, provided that he can demonstrate that he has used reasonable care and skill in trying to achieve success. (And in the normal way of things 'reasonable care and skill' will be the level of care and skill exercised by the average practitioner, in other words, not necessarily a particularly high level.)

9.5 Sale contracts and contracts for the transfer of ownership of goods

As we have already noted above, many contracts which are not properly contracts of sale can transfer the ownership of goods. We have already mentioned the building/installation contract.

Another example is a contract to service/repair a motor vehicle, when the new lubricating grease and oil or replacement parts supplied by the garage will become the property of the owner by being fitted or added to his vehicle (and the old lubricating oil drained from the vehicle will become the property of the garage).

This distinction between different types of contract caused problems in the past, because there was often doubt as to whether the obligations of the parties laid down by the Sale of Goods Act in relation to sale contracts would also apply to non-sale contracts that had the same result. This uncertainty was eventually settled by the Supply of Goods and Services Act. This provides that similar implied terms to those set out in the Sale of Goods Act for sale contracts should also apply to other contracts that result in the transfer of the ownership of goods.

9.6 The sale of goods framework

Introduction

Some of the legal jargon relating to goods can be confusing. 'Specific goods' means simply identified items, such as a particular machine, or piece of furniture or picture. 'Existing goods' means goods that exist and are identified at the time the contract is made, such as a batch of material stored at a particular location. 'Future goods' means goods that still have to be manufactured or purchased by the seller at the time the contract is made. 'Unascertained goods' means that the goods exist but have not yet been identified or allocated to the contract, for example, where there is a contract to supply 50 tonnes of oil from a tank containing 100 tonnes.

The 1893 Sale of Goods Act was the work of a great parliamentary draftsman, Chalmers, who set out to reduce the precedent law relating to contracts of sale that had developed during the second half of the nineteenth century into a single coherent set of principles. These generally reflected previous decisions, but with some improvements of Chalmers' own added. The Act then remained in force for over 85 years before it was replaced by the current 1979 Act, which made few changes of any substance and has itself now been in force, with some legislative upgrading, for nearly 40 years. Upgrading has mainly consisted of extending the provisions of the Act from sale to analogous contracts and improving upon quality and inspection provisions.

In broad terms, therefore, the legislative structure has remained unchanged for well over 125 years.

The preliminaries: sections 1–4

These sections set the scene. The contract may be made in all the normal ways, and the rules for capacity and agency are as set by the general law of contract. The contract may be for an immediate sale or for a sale in the future.

The contract may be 'conditional' as well as fully binding: 'If I can obtain the necessary finance then we have a deal', for instance.

The goods: sections 5–7 and section 16

The contract may be for the sale of existing or future goods. There may also be a contract where there is uncertainty about whether future goods can be found by the seller: 'If I can obtain the necessary import licence then we have a deal', for instance.

Where the contract is for the sale of specific goods and those goods have ceased to exist before the contract is made, or cease to exist after the contract but before delivery, then the contract is void.

Finally, as ownership of unascertained goods cannot pass until the goods have been properly identified and allocated to the contract, we cannot have a contract of sale relating to them, only a contract to sell them (in the future).

The seller has to provide goods that comply with the contract when delivery is due. But the goods do not usually need to exist when the contract is made, except where it is for the sale of specific goods. But what if those goods have ceased to exist or have been substantially damaged or are lost or damaged before delivery?

The normal position in respect of specific goods that have ceased to exist through no fault of either party is that the contract is void, so that any money paid has to be returned, and neither party has any other liability to the other (section 6). The position is the same where the goods cease to exist between contract and delivery (section 7).

The case usually used as an example is –

Couturier v. *Hastie* (1856) H sold to C a consignment of corn currently being shipped to the UK. Unbeknown to either party the goods had already been damaged by seawater and jettisoned by the ship at the date of contract. The court held that there was no contract. It is difficult, as is often the case in old cases, to be sure of the precise basis of the decision. But it was either that the contract was void for mistake – the traditional view (see below) – or that the contract could not be carried out if the goods did not exist – a more recent academic view.

But the seller may be liable to the buyer if he has, expressly or by implication, promised the buyer that the goods do exist at the time of contract. This can be where the contract is for the sale of specific goods, or for unascertained goods to be allocated from a specific source, such as an amount to be allocated from a cargo currently being carried by a named ship.

See the following case –

McRae v. *The Commonwealth Disposals Commission* (1950) This was an Australian case but is accepted as an accurate statement of the UK position. The CDC was responsible for the sale of surplus/scrapped Australian government stores. It advertised a Second World War tanker for sale, which was described as being wrecked on a reef (with co-ordinates supplied) off the coast of Papua. M bought the tanker and then organised a salvage vessel to recover whatever was possible from the wreck. When the vessel arrived it was discovered that there was no wreck at the location given by the CDC. In fact, there was not even a reef there. M sued the CDC for damages and the court decided that in advertising that the wreck was to be sold the CDC had in fact promised that it did exist (and that it was only because the CDC did so that M and other companies were prepared to offer to buy). Therefore, even though it would not normally be the case that the seller would be held responsible if specific goods did not exist at the date of contract, as the CDC had specifically stated to M that the wreck did exist in order to induce M to offer to buy it, CDC was liable to M for the price that M had paid, plus the costs of M's abortive salvage expedition.

Another example of the problem, but in a different context, is the following case –

Associated Japanese Bank (International) Ltd v. *Credit du Nord SA* (1988)
AJB had entered into a contract with B to buy and lease back to him four textile-packaging machines, supposedly installed in his factory, for just over £1 million. B was fraudulent and the machines did not in fact exist. AJB required B to give security for the money, and B persuaded CDN to guarantee the debt. (Neither bank actually checked that the machines did exist.) Then B defaulted and disappeared with the money. CDN refused payment under the guarantee and AJB sued. The court held that the existence of the machines was fundamental to CDN. If the guarantee was called, CDN would have no means of recovering their money except by enforcing a sale of the machines. As they did not exist, AJB and CDN were both labouring under a common mistake that was fundamental to the transaction and therefore the guarantee was void.

The position of B, the 'seller' under the purchase and lease back contract with AJB, would be a very different matter. B had clearly promised AJB that the four machines did exist. Had that contract come before the court then there is no doubt that the court would have made the same decision as that in *McRae*'s case. But what the court really had to decide was which of two innocent financial institutions had to stand the loss caused by a fraudulent third party, and the court decided that this should be AJB. Perhaps the court felt that AJB had more reason to check both that the machines existed and also that they were worth the £1 million that AJB paid out. The case does raise some problems, but it also illustrates the principle that where both parties to a contract relating to specific goods have a common but mistaken belief that the goods exist, and where the existence of those specific goods is of fundamental importance to the contract, then the contract is void.

The price: sections 8–9

The price for the goods may be decided in a number of different ways, fixed by the contract, fixed by a method or formula agreed in the contract, fixed by the course of dealing between the parties, or fixed by the valuation of the goods by an agreed third party. Then, as a fallback position if all else fails, the buyer must pay a reasonable price (which depends on the circumstances).

The one problem with these sections is how a reasonable price is to be fixed, if the need arises, which is rare. Hopefully this would be by agreement between the parties, to be based upon the going market rate. But if there is no machinery stated in the contract, and the parties do not agree, it might need to be done by an arbitrator/judge on the basis of expert evidence, which would be an expensive procedure to follow.

This is in fact a straightforward statement that simply reflects how most contracts operate in practice. In the vast majority of cases of course the contract will fix the price to be paid for the goods. Alternatively the price may be fixed by reference to rates or prices pro rata to the amount of effort needed from the seller to supply or manufacture the goods, or by measuring or weighing the goods to determine their actual amount, or sampling the goods to determine value, and so on. Sometimes this may be done by the parties themselves. Sometimes it will be done by a third party, as in the 'look–see arbitrations' common in years past in various import and export trades, when an 'arbitrator/expert' would simply examine the goods as they were unloaded from the ship and then fix their value and therefore the price to be paid by the buyer. Finally, there is the catch-all provision in subsections 2–3 that if there is a price to be agreed, but there is no method of arriving at that price stated in the contract, then the buyer must pay a reasonable price. The usual route to agreeing a 'reasonable' price in the commercial world is by third-party valuation by an agreed valuer, or alternates in the event that the nominated valuer fails to act.

The method of payment

Payment may be made directly by the buyer or by a third party, and it may be absolute/unconditional or conditional.

The normal method of unconditional payment foreseen by the Act was payment in cash. Payment by any other means was seen as merely conditional, allowing the seller to claim against the buyer and even to repossess the goods in the event that the payment failed; see below.

When goods are purchased through a credit or direct debit card the position is that a third party, the issuer of the card, will now have a contractual obligation to make payment in cash to the seller in due course. It is not immediate payment in cash in any sense of the word, but it has almost the same effect. The position is created by two contracts. The first is a contract between the buyer and the issuer of the card, under which the issuer agrees to pay money to each and every seller from whom the buyer purchases goods using the card, and the buyer agrees to reimburse the issuer for those payments. The second is between the seller and the issuer of the card, under which the seller agrees to accept payment from the issuer for its goods by buyers using the issuer's cards without any further recourse to those buyers if payment by the issuer fails. It would therefore appear to be the case that payment by credit/direct debit card is a contractual absolute payment, though not an immediate payment.

The typical methods of conditional payment are by cheque, bill of exchange or promissory note. Here the seller agrees to accept payment from the third party, usually the buyer's bank, in due course. However, there is no agreement between that third party and the seller, so that the seller automatically retains the right to claim from the buyer in the event that he does not receive payment in due course.

Time: section 10

If the contract is silent, the times for payment stated in it are not of the essence of the contract. The contract decides whether other times are of the essence or not.

(And 'month' means calendar month unless otherwise stated.)

The principal time promise is almost always associated with delivery/completion.

The law has always recognised the time promise as a condition, so that breach may lead to damages and/or termination of the contract or both.

Fixed dates

Time is an area of the law where there are different rules for the commercial contract and the consumer contract.

In the consumer contract the law is that any times stated for performance are merely estimated, requiring only performance within a reasonable time, *unless it is absolutely clear* that the parties intend a time stated in the contract to be of the essence.

In the majority of commercial contracts on the other hand, unless otherwise stated in the contract, time will be presumed to be of the essence, *at least as far as the principal delivery/completion promise is concerned*. 'Time of the essence' is a jargon phrase. It is simply describes the situation where the time obligation is a set period, date or time, and is a condition of the contract. The result is that any lateness, however small, without lawful excuse is automatically a breach of condition, allowing claims for damages, plus termination.

Union Eagle Ltd v. *Golden Achievement Ltd* (1997) UE was buying a luxury flat in Hong Kong from GA. The agreement required the payment of a 10% deposit, which was duly paid. The agreement then fixed the time and place for completion of the sale as 1700 hours at the offices of the seller's solicitor on a particular day. The clause then stated that any failure to comply with any of the terms and conditions of the contract would result in the deposit being forfeited as liquidated damages and the seller then having the right to cancel the sale. Immediately after contract the property market in Hong Kong soared, so that the flat was now worth very much more. The messenger bringing the payment from UE did not arrive by 1700 hours. At 1700 hours GA's solicitor telephoned UE and warned UE that her client reserved the right to rescind the contract. At 1710 hours the messenger arrived with the payment. The solicitor refused to accept the payment and at 1711 telephoned UE to confirm that GA had terminated the contract. The court decided that the agreement was commercial in nature. Therefore, once 5 pm had passed UE was in breach of the contract and as a result GA was entitled to refuse to accept the payment. It had the option to accept payment after 5 pm should it wish to do so but could not be required to accept it.

The case is a perfect example of the kind of procedure to be followed in the event of termination for lateness.

Other dates or times stated in the contract will not normally be classed as conditions unless stated. After all, in a complex contract both sides may have a number of time-related obligations to perform during the contract – giving notices, giving approval or information, delivering documents, making available facilities and so on. The law presumes that it is not normally the case in a complex contract that the parties will intend that all these interim obligations must be completed strictly to time. Many may relate to only minor aspects of the work, or may have little or no effect on the overall outcome of the contract. Therefore, the law will always consider each time obligation individually, asking what the parties intended that obligation to be. It is not yet clear whether the courts would treat interim time obligations as contract conditions, warranties, or, more likely, as innominate terms.

But any statement in the commercial contract that a particular time is to be of the essence of the contract, or that it is of considerable importance, or must be strictly complied with, or that reserves the right to terminate for lateness, and so on, or even a contract that simply states a fixed time for delivery or completion, will normally mean that the obligation is of the essence.

A textbook example of the strict application of the doctrine, which raised some interesting side issues, was the following –

> *Bowes* v. *Shand* (1877) The case concerned two contracts for the sale of 600 tonnes of rice, 'to be shipped at Madras during March and/or April'. Most of the rice was actually loaded into the ship during the month of February, with only a few hundredweight being loaded during the first week of March. The ship then sailed for the UK some three weeks later. The bills of lading covered rice loaded before and during the contract period. When the rice arrived in the UK the buyer refused to accept it. The court held that the words of the contract must decide, and they were clear. 'Shipped' meant loaded into the ship, not the date of sailing of the ship, and 'March and/or April' meant that loading must both start and be completed within that period, which was clearly not the case. Rice loaded during February was not rice loaded during March. Therefore the buyer had no obligation to accept the goods if he did not wish to do so.

This case is unusual because the usual argument is about goods delivered late rather than goods delivered early. We do not know why the buyer was unwilling to accept the goods. Most probably the reason was that his contract to resell the rice was by description, the description being 'rice shipped in India during March and April', and as he would only be able to produce bills of lading for rice shipped during February he was unable to provide goods that met that description. Another interesting point is that the court actually said that, where appropriate, time could form part of the description of the goods and that therefore not only were the goods delivered outside the contract period, but they were also not the goods that the seller had contracted to deliver.

Another example of the perils of the time of the essence principle, this time relating to payment, is the following –

> *Scandinavian Trading Tanker Co AB* v. *Flota Petrolera Ecuatoriana* (1983) This case concerned a contract to hire a ship. The hire charge was payable monthly in advance and the payment clause then specifically stated that if any payment was not made on time the owner, ST, could cancel the contract. When the contract still had a year to run FPE was 48 hours late in making a payment because of an administrative muddle with its bank. In the meantime the hiring rates for ships had risen steeply. ST refused to accept the payment and immediately terminated the contract (and then, because FPE had no choice at such short notice, rehired the ship to FPE at a much higher monthly rate). Both the Court of Appeal and the Lords unanimously rejected the claim by FPE to set aside ST's actions. Their decision was that FPE had gone into the contract with its eyes wide open. If FPE did not like

the payment clause it had two alternatives, to ask for the clause to be changed, or to give clear instructions to its bank that on no account were payments to be delayed. FPE had done neither of these, therefore it must accept the risk.

There are two main exceptions to the strict time of the essence principle in commercial contracts.

The first is non-trading contracts. In particular in relation to agreements relating to trusts, company administration and land (leases, sales and so on), time obligations will not normally be of the essence but simple estimates of a reasonable period. See *United Scientific Holdings* v. *Burnley B C* (1978) dealing with the periods stated in a lease for rent review procedures. This is also the case for purely relationship contracts, such as employment, consultancy, agency, consortia/partnership agreements, and so on.

The second exception is building contracts, perhaps due to the longer timescales involved. For example, in *Ampurius Nu Homes Holdings Ltd* v. *Telford Homes (Creekside) Ltd* (2013) a delay of almost seven months to the completion of a building contract by the builder was held not to be sufficient to justify termination by the client. The court held that as the building was then to be leased by the client to a third party on the basis of a 999-year lease, the delay had not lost the client any significant amount.

However, even here if an agreement is commercial and states clearly that time of performance is to be treated as fixed, then time will be of the essence. See *Union Eagle Ltd* v. *Golden Achievement Ltd* above.

Estimated/approximate dates

The alternative to a fixed date is an estimated date, or approximate date.

The law is straightforward. Where the contract does not state any fixed time to carry out any obligation it must be carried out within a reasonable period. If the contract states that an obligation is to be carried out within an estimated period of time then it must be carried out within a reasonable period before or after that.

If the obligation is then carried out within that reasonable period, the party concerned will have complied with the contract. If however he fails to carry out that obligation within a reasonable time he will be in breach of condition.

What is reasonable will depend upon the circumstances.

There are few guidelines to what is reasonable. Most important is the intention of the parties, as shown by the contract and the factual matrix that lies behind it, or perhaps as shown by the course of previous dealings between them. There is also the custom within the particular trade or profession concerned.

Apart from that there is very little. There is no guidance from court decisions. Therefore, it is very difficult to know when the innocent party is properly entitled to terminate, if he should actually wish to do so. The only suggestion that one can make is to be cautious. But as a very rough rule of thumb, anything over 15–20% beyond the estimated period will be too long.

Managing lateness

The *Union Eagle* case is a perfect example how a party might act if the other side is in breach of a fixed-time clause (and indeed of any contract condition) if he wishes to enforce immediate termination of the contract for that breach.

Very often the buyer will not want to lose his contract, but want the seller to proceed. Better a late supplier than having to repeat the whole procurement process again from scratch. If this is so, the buyer will usually contact the seller, instructing or asking him to deliver/complete or to forecast when he will complete and so on. By doing so the buyer is agreeing to waive his right to immediate termination (see section 11(2) of the Sale of Goods Act). So the seller is now required to deliver/complete within a reasonable period. Should the seller do so then of course all is well. Should the seller fail, the buyer is entitled to reassert his right to terminate by giving the seller notice that delivery must be completed within a further reasonable period. What period will be reasonable will again depend on the situation at the time, in particular the amount of work that the seller has to do to be able to deliver, and likely delays for reasons outside the seller's control. If the seller then fails to complete within that period, the buyer can terminate the contract. For a good example of how this can operate in practice see the following.

> *Rickards* v. *Oppenheim* (1950) O ordered a Rolls-Royce from R. R was responsible for providing the chassis with a specialist coachbuilder acting as a subcontractor to build the rest of the car. The car was originally due for delivery in March 1948, and the contract specified this as a fixed delivery date. However, the subcontractor was late. O did not cancel but pressed for delivery in time for the Ascot race meeting in May. The car was still late. With the consent of R, O then spoke directly to the coachbuilder. He promised O that he could definitely have the car ready for delivery by 12 July. O was due to travel to France on 3 August, and wanted to take the car with him. Having a promise that the car would be built by 12 July he therefore wrote to R saying that he could accept delivery of the car up to 25 July (which allowed him reasonable time to make arrangements to take the car to France a week later, and also gave R two weeks after the period promised by the subcontractor to prepare and deliver the car), but that he would not accept the car if it was delivered after that date. The car was not ready on 25 July. In fact, it was not ready for delivery until October. When it was offered to him in October O refused to accept it. The court decided that O was fully entitled to do so.

On the other hand, an example of how not to proceed is given by the following case –

> *Hartley* v. *Hyams* (1920) This was a contract for the delivery of 'processed doubled yarn' at the rate of 1100 pounds per week over a ten-week period in 1918 from the start of September to 15 November. Hartley actually commenced delivery on 26 October with the delivery of 550 pounds of yarn and then delivered seven

further quantities, all over 500 pounds, over the period to February 1919. (This was an appalling delivery record – the problem was that a subcontractor was having difficulty with the doubling process and serious delays were being caused by the Cotton Control Board.) During this period Hyams wrote several times to Hartley asking for better deliveries but did not mention the possibility of termination. Then on 13 March, when in fact Hartley was ready to deliver the remainder of the yarn, Hyams terminated without giving any further notice. The court held that by writing to request better deliveries after 15 November, Hyams had in fact waived the right to terminate, and was therefore not able to do so without giving reasonable notice.

Transfer of ownership and 'quiet possession': section 12

It will be an implied condition of the contract that the seller will have the right to sell the goods at the appropriate time, whether he is the actual owner or not, and that he can also give 'quiet possession' of them to the buyer. The goods must also be free from any legal claims, unless these have been disclosed. Where the seller is not selling full ownership of the goods, the seller must disclose all the legal claims of which he is aware before contract and not allow any claims against the goods by himself or the previous owner.

The object of the sale contract is to transfer ownership of the goods from the previous owner to the buyer. Any contract of sale will be for the transfer of full ownership, unless, which is rare but not unknown, it is clear that both seller and buyer knew that the seller did not or might not have full title, but was selling the title that he had.

Usually the seller will be the owner. Occasionally someone else can be the seller. Examples are an agent in possession and authorised by the owner to sell on his behalf, a liquidator or administrator of a company, or a bailiff ordered by a court to take possession of the goods of a debtor and sell them in order to recover money for a creditor. Another may be when goods are exposed for sale and then sold at an ancient 'market ouvert'.

There are three important considerations. They are 'ownership' – which of the parties will own the goods at any specific moment, 'possession' – which of the parties will hold the goods at any specific moment, and 'quiet possession' – will the buyer be able to have the right to use the goods without hindrance.

(There is also 'risk' – which of the parties will be responsible if the goods are damaged or destroyed. The law of sale is not concerned with risk directly, but assumes that risk will transfer with ownership unless the contract states to the contrary.)

Two cases relating to second-hand cars, which produced odd results, serve to illustrate the problems of ownership. Both were governed by section 12 of the 1893 Sale of Goods Act, which was the same as the current Act –

Rowland v. *Divall* (1923) A stolen car was sold to D, who was completely innocent, and then resold by D to R who was also completely innocent. R used the car for a short period, then sold it on to someone else. Four months later the true facts came to light. The car now had to be returned to the real owner. R unscrambled his resale contract at a small profit and then claimed his original purchase price back from D. D counterclaimed on the basis that R's claim should be reduced, because he had had the use of the car for several months. The court's decision was that D had to have the right to sell the car. Clearly he did not. Although he was blameless, he could not give ownership or quiet possession. In fact, R got nothing from the contract, and the fact that he had the benefit of using nothing for four months and then made a profit out of selling nothing and buying it back again did not change that. D was therefore liable to repay the whole of the original price and could not counterclaim anything.

Karflex v. *Poole* (1933) Karflex provided P with hire-purchase finance on a second-hand car that had been sold to him by a third party, King, who actually had obtained possession of it from the rightful owner by fraud. P paid the down payment and then defaulted. Karflex reclaimed the car and sued P for the remainder of the hire-purchase charges. Only then did the true facts emerge. The car was now claimed by the true owner, and Karflex tried to correct the situation by buying the car from him. However, P now claimed that as Karflex had not owned the car when the hire-purchase contract was signed, Karflex could not give him any right to hire the car, or the option to purchase it, and was therefore in breach. P claimed therefore that he had no liability for the hire-purchase charges and was entitled to the repayment of his deposit. His claim was accepted. He recovered his deposit in full.

The perfect example of the problems of quiet possession comes from one of the classic cases relating to trademarks, again under the 1893 Act –

Niblett v. *Confectioners' Materials Co Ltd* (1921) N, a grocery wholesaler, bought a large quantity of tins of condensed milk from CMC, to be shipped from the US. When the milk arrived in London, N discovered that the majority of the tins carried the name 'NISSLY' on the labels. Consignments of condensed milk carrying this trademark had already been imported into the UK by others, and the NESTLE company had already obtained an injunction against the other importers banning the resale of the milk, because 'NISSLY' infringed the 'NESTLE' trademark. N realised immediately that it would be hopeless for him to try to sell the milk without removing the labels from the tins, which was both expensive and also greatly reduced the price at which he was able to resell. (There was also no commercial sense in a wholesaler of dairy products getting into a dispute with a major producer of dairy products.) N therefore removed the labels and sold the tins at a significant loss, then sued CMC for damages, and won. The court held that CMC clearly could not give N quiet possession. The court also held that CMC did not have the right to sell the goods and, further, that they were not (see below) of merchantable quality, because the court felt that they were not saleable.

Delivery – the transfer of ownership/possession

It is sometimes said that the seller must deliver the goods to the buyer. This is not correct. Actually the seller must make the goods available for collection by the buyer at the point of delivery. It is then the responsibility of the buyer to take possession. In other words, what needs to take place is a transfer of ownership, not any physical movement of the goods. (The practical consequences are, for instance, that unless specified otherwise, where goods are delivered 'ex works', loading those goods on to the vehicle is at the cost and risk of the buyer.)

There are a number of different ways in which delivery can take place.

Most commonly the seller will place the goods at the delivery point and allow the buyer to remove them. After all, this is what happens in almost every retail sale in a shop or supermarket.

It may take also place by the seller transferring to the buyer the means of control of the goods. An example of this is a car dealer giving the buyer the car logbook and ignition key, or a seller giving a buyer access to the place where the goods are located.

Where the goods are in the possession of a third party, delivery may take place in other ways. The seller may instruct the holder that the goods are now the property of the buyer, or the holder may confirm to the buyer that the goods are now available to the buyer. The seller may give the buyer an order to the holder to release the goods to the buyer. Then delivery may take place by the seller giving to the buyer the 'document of title' to the goods. 'Document of title' has a very special meaning in law – it is a document that 'represents' the goods, such that handing over the document automatically transfers the ownership of the goods. In fact, the only document that has this special status under English law in relation to goods is a bill of lading.

It is an underlying condition of the contract that the buyer must take delivery of the goods once the seller has offered to him goods that comply with the contract.

When the goods are perishable in nature, or the place of delivery is the buyer's own premises, then the duty to take delivery is of the essence of the contract. Further, where the contract is for the immediate delivery of the goods, as in the normal consumer sale or what is sometimes called a 'spot contract', the buyer also will have an obligation to take immediate delivery. In other cases, where effectively the buyer simply has to take the goods away, then his obligation is merely to do so within a reasonable time.

The buyer is not under any obligation to accept part deliveries of the goods, unless the contract permits delivery in instalments. Where, however, delivery in instalments is allowed, then the buyer is required to accept delivery in this manner.

Delay in accepting delivery is one thing. Rejection or the refusal to accept delivery, however, for any reason other than a failure by the seller to comply with the requirements of the contract, will be a breach of the contract, usually justifying the seller in terminating the contract.

Transfer of ownership, as opposed to delivery: sections 16–18

Ownership cannot be transferred until the specific items to be sold have been properly identified. Ownership will be transferred when intended by the parties.

Unless the parties intend otherwise –

- where goods are sold in a fit condition to be delivered, ownership will pass when the contract is made;
- if something needs to be done to get the goods ready to deliver, ownership will pass when the goods are ready and the buyer has been notified;
- when the goods need to be measured/weighed etc. to decide the price, ownership will pass when this has been done and the buyer has been notified;
- when the goods are being sold on a 'sale or return' basis, ownership will pass when the buyer accepts the goods or does not return them in due time;
- finally, where the contract is for the sale by description of future or unascertained goods, ownership will pass when goods in a deliverable state have been appropriated to the contract with the parties' agreement. This may be by delivery to another to deliver them to the buyer, or where the goods can be clearly identified.

In the great majority of commercial contracts the provisions of the contract will decide when ownership is transferred from seller to buyer.

As regards specific goods, when the goods are in a deliverable state at the time of contract, then ownership will be transferred to the buyer immediately. As an example, a model car in a toy shop will become the property of the buyer immediately the contract is made. If, however, something needs to be done to put the goods into a deliverable state, say a car in a dealer's showroom needs to be taxed and checked before it is ready to be driven on the roads, then ownership will not pass to the buyer until the seller has done whatever is necessary to put the goods into a deliverable state and has informed the buyer that this has been done.

The 'quality' of goods: sections 13–15

These sections lay down a number of implied conditions. First, the goods must comply with any description included within the contract. Second, if the goods are sold by sample then the bulk of the goods must comply with the sample as well as the description. Finally, where the goods are sold in the course of a business, the goods must be of satisfactory quality and fit for use for any particular purpose which the buyer has notified to the seller.

This is an area that has always been the focus of much interest and national policy considerations, both in terms of legislation and of litigation – legislation to protect consumers and litigation arising out of customer dissatisfaction, both merited and unmerited.

Essentially the law divides the quality of any goods into three separate areas: their durability/quality of manufacture or finish, their usefulness, and their description or specification. They are all contract conditions, implied into every contract by the Sale of Goods Act or the Supply of Goods and Services Act. However, the law treats the conditions differently. The parties are permitted, subject to certain very major qualifications, to limit or even exclude liability for breach of the implied conditions of quality and usefulness, provided that the exclusion/limitation is 'fair', within the definitions laid down in the Unfair Contract Terms Act 1977. The implied conditions of specification/description are almost absolute.

The implied condition of description
The implied conditions are contained in sections 13 (description) and 15 (sample).

1. Where there is a contract for the sale of goods by description, there is an implied condition that the goods will correspond with the description.
2. If the sale is by sample the goods must correspond with the sample.
3. If the sale is by sample as well as by description it is not sufficient that the bulk of the goods correspond with the sample if the goods do not also correspond with the description.

Section 13 applies to all contracts under which goods are sold or supplied, not just to contracts made in the course of a business. Almost everything sold or supplied under the commercial contract will be by description. This will also be the case under the vast majority of consumer contracts, and a great proportion of sales between private persons. The condition is an absolute requirement and cannot be excluded or limited except where it is clear from the contract that, for instance, a seller is only accepting the obligation to try to meet a particular description, which is very rare indeed, or is selling specific goods, such as an antique, without any firm undertaking that it actually is what either or both of the parties believe it to be.

Two examples, out of many –

Leaf v. *International Galleries* (1950) L bought a watercolour painting from IG. It was described by IG as being by Constable. Five years later L tried to sell the painting and was then told that it was not a Constable. He applied for rescission, that is, cancellation of the contract. The court held that section 13 (under the 1893 Act) applied, so that the sale of the painting was by description, but that as L had owned the picture for five years before discovering the breach it was too late for him to ask for rescission. The court actually applied section 35 of the Act, which provided that once the buyer has had reasonable time to examine his purchase then he must accept or reject, and if he does not reject it then he loses the right to reject.

> *Beale* v. *Taylor* (1967) B bought a second-hand car from a dealer, T, described as a 1961 Triumph Herald 1200 cc. It then emerged that the rear end of the car fitted the description, but that the front end was that of a 1948 model. The court held that this was a breach of section 13. B was entitled to cancel the contract and recover the price that he had paid.

The reasoning behind section 13 is straightforward. As explained in one famous judgment, to supply beans when the contract is for peas has got to be a failure to carry out the contract. The buyer must be given what he contracted to buy, and no law could allow the seller or supplier to get away with anything less. But there are three related but distinct questions that must be addressed.

First, what will constitute the 'description' of the goods? In a supermarket the answer is simple. The description of the goods will be what is said on the packet. (Though even then there may be some degree of uncertainty. There may be lists of ingredients without the exact proportions stated, or tables of 'typical values'.) Commodities will have comparatively simple descriptions. So can some equipment. In *Varley* v. *Whipp* (1900) the contract was simply for the sale of 'a nearly new reaping machine' (it wasn't). But with complex manufactured equipment, such as an airliner or an offshore oil production platform, then the contract will automatically need to include a very large, complex and detailed specification of what is to be supplied. That specification is the description of what is to be supplied. But is every single statement made in that specification a part of the description, so that any breach will justify termination? (After all, in the natural way of things it is highly unlikely that what is supplied can ever comply in every single tiny detail with the specification.)

The earlier cases would certainly have implied that this was so. Once it was clear precisely what the contract description was the law required that description to be complied with – absolutely. Therefore any measurable discrepancy, however slight, would be enough to constitute a breach of the condition and justify termination of the contract. Perhaps the clearest example of this approach in its most rigorous form is that of *Arcos Ltd* v. *E A Ronaasen & Son* (1933), the barrel staves case referred to above, in which the buyer was held to be entitled to terminate the contract even though the court found that commercially (as opposed to legally) the goods were acceptable against their specification and of saleable quality.

In more recent cases the courts have relaxed their approach. In the *Reardon Smith Line* case mentioned in Chapter 8, the court held that the description of the ship that was to be built only included the substantial elements of what was to be supplied, as opposed to minor details. Again, in the case of *Ashington Piggeries* v. *Christopher Hill Ltd* (see below), the law was stated as follows –

> The description is … confined to those words in the contract which were intended … to identify the kind of goods that were to be supplied. It is open to the parties to use a description as broad or as narrow as they choose. But ultimately the test is whether the buyer could fairly and reasonably refuse to accept the physical goods proffered to him on the ground that their failure to correspond with what was said about them makes them goods of a different kind from those he had agreed to buy.

This is clearly a much less rigid position than that suggested by the early cases. But even if it is not a breach of the section 13 condition, a failure to comply with the specification will still be a breach of either an innominate term or a warranty. Essentially, the rule for the seller/supplier must always be that he fails to comply with the specification at his peril. The same is also true for the buyer. An example is *Frederick E Rose (London) Ltd* v. *William H Pim Jnr & Co Ltd* (1953) in which a simple mistake led to a buyer thinking that 'horse beans' and 'feveroles' were interchangeable terms. They were not. But having bought horse beans by description he had no choice but to accept the consequences of his mistake.

The next question is what will actually constitute a breach. Clearly, any serious or material default will be a breach, but what about a small breach? There is a well-known and often mistranslated Latin tag – *de minimis non curat lex*, the law will not bother with every little detail (or breaches of description). The argument is often raised but is seldom successful.

Where a contract depends on a complex specification the court will always take the view that if the parties are prepared to agree upon a precise and detailed description of the goods, then they will have done so because the details are important. If, in addition, the two parties are in dispute, then it is only even more likely that the detail matters. Therefore the seller would only succeed with a *de minimis* argument if he can prove that the breach really is totally trivial in nature, which in practice is extremely difficult.

Where the contract description is simple the problem may be easier to solve. Contrast the two following cases.

> In *Wilensko STD* v. *Fenwick & Co* (1938) 1500 pit-props were supplied against detailed sizes/lengths, with a 10% limit, plus or minus, on certain sizes. On the first consignment some of the props were rather shorter than they were supposed to be. In addition, the 10% limit on one size of prop was clearly exceeded, but by only a small amount. The goods were of good quality and perfectly saleable. The court clearly felt that two breaches were too many and allowed termination for breach.

> *Shipton Anderson & Co Ltd* v. *Weil Bros* (1912) Under a contract for the delivery of 4500 tons of wheat plus or minus 10%, S delivered 4950 tons plus 55 pounds. W rejected the whole delivery, even though S did not charge for the 55 pounds. The court held that as the excess was about 1 pound per 100 tons, this really had

to be a *de minimis* situation, and therefore W was ordered to pay for the goods, or in fact compensate S for the loss that he had incurred in selling the goods to someone else.

There third question of course is what practical remedies there are for a buyer who has been offered goods that do not comply with the contract description. Of course he may terminate, but termination is often not a very practicable commercial option, especially when dealing with contracts for the purchase of long-delivery or capital equipment. He may also be able to claim damages, but damages, which are discussed in detail in Chapter 16 below, present their own problems as well. For these reasons, contracts often concentrate on commercial remedies for the problem – inspection, testing, quality control, defect repair, etc. – on the principle that it is far better to correct the problem than bother the law.

The implied conditions of satisfactory quality and fitness for purpose
Section 14, as amended by the Sale and Supply of Goods Act 1994, sets out two implied conditions that apply to all sales of goods in the course of a business.

The first condition is that the goods must be suitable for use for any purpose of which the buyer has informed the seller prior to or at the time of the contract and where the buyer is at least to some extent relying on the skill of the seller to supply compliant goods.

The second is that the goods must be of satisfactory quality. To be of satisfactory quality the goods must meet the standard that a reasonable person would regard as satisfactory taking account of the description of the goods and 'all relevant circumstances'. Among the relevant circumstances are that the goods must be fit for use for all purposes for which they are usually used, that they must be of satisfactory appearance and finish, free from minor defects and safe to use, and have a reasonable operating life. There are the normal qualifications relating to defects that would be apparent on inspection by the buyer or of which he has been informed and so on.

The implied conditions will form part of any contract, unless they have been excluded clearly. See for instance *Dalmare SpA* v. *Union Maritime Ltd* (2013) in which a contract for the sale of a ship 'as she was at the time of inspection' was held to include the conditions.

The concept of 'fitness for purpose' does not present too many problems. The concept of 'satisfactory quality' presents far more, and therefore an explanation of the development of the current position may be helpful.

The early cases did not concern themselves with the quality of manufactured goods and as a result did not produce any particular theory as to what standard of quality of the goods should be expected from the seller. But in section 14 of the original 1893 Act, Chalmers seems to have almost single-handedly produced the concept of the 'merchantable' quality of goods. Very loosely, the term meant 'buyable', recognised by a buyer as something that he would be willing to buy against that description. The Act gave no help in defining what might or might not be merchantable quality so the matter was left to the courts. Over the succeeding period the courts then developed the doctrine that to be of merchantable quality goods must be –

- free of defects – but only on the date that they were delivered, and
- capable of being used for one of the purposes for which goods of that kind might normally be used.

Grant v. *Australian Knitting Mills* (1936) G bought a pair of 'pre-shrunk' underpants. Pre-shrinking cloth involved dipping it in a very strong alkaline solution. Unfortunately, the cloth from which his underpants were made had not been properly washed afterwards, with the result that G suffered severe dermatitis. The defendant claimed that the underpants were free of defects and manufactured to a proper standard. It was just unfortunate that they had not been washed. The court listened patiently to this defence, and then simply declared that, if G suffered considerable pain and distress as a result of wearing them, the underpants were neither defect-free nor of merchantable quality.

But there were obvious shortcomings with this definition. It was aimed at the lowest common denominator. Of course it was completely adequate for simple purchases, but left much to be desired when applied to complex manufactured items. (It was, however, repeated in the 1979 Act.)

To begin with, it did not, in any sense of the word, actually attempt to define 'quality' at all, whether of raw materials, manufacture, finish or 'value for money'. It made no attempt to address the question of reliability of use or operation. Finally, it signally failed to deal with the question of items that could be used by a buyer for a number of different purposes. The first attempt to deal with these shortcomings was made by the Sale and Supply of Goods (Implied Terms) Act 1973, which covered a range of non-sale contracts, especially contracts for hire purchase. Then the traditional wording of section 14 was amended by the Sale and Supply of Goods Act 1994 to produce the current law.

The implied condition of satisfactory quality

The condition applies to all goods sold in the course of a business, both goods of the kind normally sold as a part of the business and goods that are sold incidental to the business.

Stevenson v. *Rogers* (1999) R was a fisherman. He had previously bought and then sold another fishing boat. He now sold his latest boat to S, and then bought another. S was dissatisfied with the boat that he had bought from R and claimed damages on the basis that the boat was not of satisfactory quality. His claim was successful. The court held that the sale by R of his fishing boat was in the course of his business as a fisherman.

The condition is not implied into private sales, but could of course be specifically included, if, for instance, a private person sold goods as a part of a paying hobby. The condition applies to both new and second-hand goods. In other words, if a company sells off second-hand machinery or a firm of solicitors sells a vehicle because they are out-dated or surplus to current requirements, that is still a sale in the course of business. Also, where goods are sold by someone who is not acting in the course of a business, but who is selling through an agent who is acting in the course of his business, then the condition will also apply, unless the buyer is aware of the status of the seller.

Satisfactory quality starts from the previous law on merchantable quality and goes on from there. There were a reasonable number of earlier cases concerned with what was merchantable and what was not. We have already seen in *Niblett* v. *Confectioners' Materials Co Ltd* above a case where goods were held not to be of merchantable quality because they infringed a trademark. Other cases have covered such things as 'Coalite' fuel – unsaleable because there was an explosive cartridge in it; a bun containing a stone; and a ginger beer bottle that shattered. All these cases would be covered by section 14.

There are very few cases in which the courts have discussed how the definition of satisfactory quality set out above should be applied to any actual case. This is not perhaps surprising, and probably does not present too many difficulties, because the definition in the Act is both very comprehensive and well expressed. The only question that the organisation really needs to ask itself is whether a reasonable man would accept that any particular goods were of satisfactory quality or not when looked at in comparison with the requirements set out in that particular contract.

It is perhaps also worth referring to two cases that give examples of the approach being adopted by the courts to the application of the definition of 'merchantable quality' to actual situations (where merchantable quality would be identical to satisfactory quality) –

B S Brown & Sons Ltd v. *Craiks Ltd* (1970) This case concerned two orders for the purchase of a large quantity of cotton/rayon cloth. The cloth was bought by C, a merchant who intended to use it for making dresses. However, C did not tell B the purpose for which he intended to use the cloth. B on the other hand sold the cloth on the basis that it was of reasonable quality for 'industrial uses', such as making into bags or overalls. Cloth for dresses needed to be of a much higher quality. The price at which the cloth was sold was slightly high for 'industrial' cloth, but very cheap for 'dress-making' cloth. C then rejected something like two miles of cloth on the basis that it was unsuitable for dress-making, and was therefore not of merchantable quality. The court had little difficulty in deciding that the contract was a sale by description and that both parties did in fact accept that the cloth met that description. The court then decided that the cloth was of merchantable quality. It was industrial cloth and was clearly usable for industrial purposes, and the price was not unreasonable for industrial cloth. Therefore, the cloth was of merchantable quality. (In fact, the court used the price to help it to decide that what was being sold was 'industrial cloth', and that therefore the cloth was of merchantable quality – one of the considerations referred to in the definition of satisfactory quality in the revised section 14(2).)

(In giving judgment the Lords quoted a comment by Lord Reid in *Hardwick Game Farm* v. *Suffolk Agricultural Poultry Producers Association* the previous year, also relating to section 14.

What subsection 2 now means by merchantable quality is that the goods in the form in which they were tendered were of no use for any purpose for which goods which complied with the description under which those goods were sold would normally be used, and hence were not saleable under that description.

This is an objective test: 'were no use for any purpose …' must mean 'would not have been used by a reasonable man for any purpose …' Again, while the references here to merchantable quality are no longer law, this is an example of the most eminent judge of his time bringing in the standard of the reasonable person to support his decision.)

M/S Aswan Engineering Establishment Company v. *Lupdine* (1987) This case concerned waterproofing material sold for export to the Middle East, or more accurately, the subcontract under which L, the manufacturer, bought the plastic buckets in which its product was to be packed. L purchased a large quantity of heavy-duty plastic pails, after first buying and trialling a small sample. The manufacturer's data sheet for the pails stated that they were capable of being stacked up to six high when full, at temperatures of 25 degrees centigrade. The pails, each holding some 20 kilos of material, were stacked up to six high in containers and shipped to Kuwait. The containers were unloaded from the ship and then left to sit on the quayside in full sunlight. The temperature inside the containers rose to the region of 70 degrees centigrade. The lower tiers of pails collapsed and split open. As a result the entire consignment was lost. Expert witnesses agreed that, given the way the pails were stacked, they were bound to collapse once the temperature exceeded 60 degrees centigrade, but that they would have survived if the temperature had not exceeded 52 degrees centigrade, and that if they had been stacked on temporary shelves inside the containers, so that the bottom layers of pails had not had to cope with the weight of several upper tiers, they would have even survived 70 degrees centigrade. Given this evidence, the court had little difficulty in finding that the pails were of merchantable quality.

As always there are several practical learning points that arise from this case. The case arose because Aswan sued Lupdine for the loss of its goods. Lupdine had failed to allow for what were actually quite predictable conditions when it packed the containers. In fact, Lupdine had no real defence to the claim, and probably could not afford to meet the damages claimed or the cost of litigation, so that when the action was commenced Lupdine brought in the supplier as a third party, and then promptly went into liquidation. The liquidator therefore had little to lose by pressing on with the claim against the supplier, even against the weight of the expert evidence, simply to try to offset the claims being made by Aswan. The case then developed into a two-pronged attack to try to recover damages from the supplier, with Aswan claiming damages in negligence on the basis that the supplier should have foreseen the loss and owed a duty to Aswan, and the liquidator claiming damages for breach of contract.

Both claims were decisively rejected.

Aswan's claim was rejected because the court found that no duty of care could exist between the supplier and Aswan (and the supplier had in any event behaved perfectly reasonably).

The claim for breach of contract on the basis that the pails were not of merchantable quality, simply because they had failed in the heat, was also decisively rejected. However good the claim might have looked at first, it was doomed to fail once the expert evidence was produced. What was even more significant was that the expert witnesses for the two sides agreed that the pails could have coped with the temperature if they had been loaded into the containers differently. In addition, the information given by the supplier in his data sheet should have alerted Lupdine to the need to check how to stack containers that were to be delivered to Kuwait. There could be no question of claiming that the pails were not fit for purpose because Lupdine had previously trialled them. The court then clearly adopted the commonsense approach of looking at what was reasonably to be expected from the supplier. He had supplied pails that matched their description, and that were capable of coping with the demands of very high temperature. He had also given a warning about how not to pack them. If the buyer did not pack them into the containers correctly that was not his fault or responsibility.

These are but two of a whole series of cases in which the courts have used much commonsense and wisdom in trying to apply the rigid doctrine of merchantable quality to real problems. It is safe to predict that the approach to problems of 'satisfactory quality' will be no different.

The implied condition of fitness for purpose

Again we are dealing with an implied condition, with strict responsibility for the seller to comply, and again the implied condition applies automatically only to business sales. The condition will apply to any contract where the buyer has made known to the seller the purpose for which he intends to use the goods, unless the seller can produce clear evidence to demonstrate that the buyer did not rely upon his skill or judgement. Reliance on skill and judgement may be partial only; see the *Cammell Laird* case below. Where the goods have only one use, the buyer does not need to make it clear that that is the purpose for which he intends to use the goods, because that will be implied automatically. Where the goods can be used for more than one purpose or where the buyer intends to use them for a non-standard use, then the buyer will need to identify the actual use for which he is buying in order for the condition to apply.

The standard of fitness required is that the goods must be 'reasonably fit' for use. They do not have to be ideally suited, or proof against misuse or incompetent operation. Nor need they be capable of dealing with any particular problem not notified to the seller (such as an allergic reaction triggered by ingredients in cosmetics or the materials from which clothing has been made).

Two examples –

Cammell Laird & Co Ltd v. *Manganese Bronze and Brass Co Ltd* (1934) CL bought propellers for two cargo ships from MBB. The contract contained a specification for the propellers written by CL and contained terms requiring the propellers to be suitable for operation with the specific diesel engines to be fitted to the ships, and to be to the satisfaction of both CL and its customer. MBB were then to manufacture the propellers using their own proprietary process, and to machine the propellers to achieve the final shape and their precise profiles. The propeller for the first ship worked perfectly. However, the second propeller was so noisy in operation that the ship was in fact refused its Lloyd's certificate. The reason why the propeller was so noisy is uncertain. It may have had something to do with the blades of the second propeller being thinner than those of the first. (Tests showed that the first [successful] propeller worked properly on both ships, so that the problem lay in the second propeller, not in the ships or engines.) MBB made a replacement propeller, but it was also far too noisy. MBB then made a second replacement propeller, which was successful. This action was mainly about which of the two companies should bear the cost of the two replacement propellers and the sea trials. All four propellers complied with the contract description (section 13 of the Sale of Goods Act) but one original and one replacement were not to the satisfaction of either CL or its customer. The court found that those terms were contract conditions and that MBB were in breach. The court then turned to the Sale of Goods Act implied terms relating to quality. The court decided that the failed propellers were of merchantable quality, because they were properly manufactured propellers and could be installed in ships. However, CL had relied upon MBB's skill and judgement in the final machining of the propellers to make them suitable for the purpose of operating with those particular ships' engines. This was a matter exclusively within the skill of MBB. Reliance by CL on MBB's skill and judgement need not be exclusive, so long as it was an inducement to CL to award the order to MBB. The implied condition of fitness for purpose did therefore apply, and MBB were in breach, because, although they would 'drive' the ship, if the propellers prevented the ship achieving Lloyd's certification, they prevented the owners from insuring the ship and therefore operating it to carry cargo.

Ashington Piggeries Ltd v. *Christopher Hill Ltd* (1972) CH was a mink farmer. He bought feed for his animals from AP, who was an experienced supplier of a wide range of animal feeds. AP supplied feed containing 'herring meal' which it had bought from a supplier in Norway. Unfortunately, the meal contained a preservative, DMNA, which is not seriously harmful to other animals but is lethal to mink even in small doses. However, the feed complied with the contract specification. CH lost the vast majority of his livestock and sued AP. The court found that as the feed met the contract specification, AP was not in breach of description. However, as AP was expert in animal feedstuffs and was being relied on by CH to supply feed suitable for feeding to mink, AP was clearly in breach for failing to supply goods that were fit for the purpose for which they were supplied. In addition, as what was being sold was described in the contract as 'mink feed', AP was also in breach for failing to supply goods of merchantable quality.

The rules about delivery and related issues: sections 16–37

These sections set out a series of basic rules that will apply unless the contract provides to the contrary.

Where the goods are ready to be delivered at the date of contract, ownership will pass to the buyer immediately and delivery will be due within a reasonable time. Where the goods are not ready for delivery at the date of contract, then ownership will pass once the goods have been identified and anything else that is needed to ready them for delivery has been done, and delivery will be due within a reasonable time.

Delivery will normally take place at the seller's place of business.

Acceptance of the goods by the buyer will take place when he confirms to the seller that he accepts them or he acts in a way that implies that he has accepted them.

Payment will be due at the same time as delivery.

The risk of damage or loss of the goods will pass to the buyer at the same moment as the transfer of ownership.

There is then a series of rules dealing with trading issues –

- the possible reservation of rights in the goods by the seller after sale (usually until he has received payment);
- rules for the transfer of ownership of goods in bulk;
- sale by persons who are not the owner or in market ouvert;
- problems concerning dealings with the goods by a person whose ownership is defective;
- deliveries by instalments, by a carrier, or of the wrong quantity.

This part of the Act sets out simple rules that cover a variety of topics that are ancillary to the main issues dealt with above. They are aimed at a small number of areas, simple (essentially consumer) sales, export/commodity trading contracts and ownership issues. They need careful reading, but do not raise many points that need comment.

The seller should deliver to the buyer the quantity of goods required by the contract. There can be of course no other position. However, what is the position if the seller does not deliver the correct quantity of goods? The basic rules are laid down in section 30, which in effect gives the buyer a series of essentially commonsense options.

If the seller delivers less than the contract quantity the buyer can reject the goods, or he may accept them, in which case he has to pay for them at the contract rate.

If the seller delivers more than the contract quantity the buyer may reject the whole, accept the whole, or accept the contract quantity and reject the rest, but he must pay for whatever he accepts at the contract rate. (This same rule also applies when the seller delivers goods to the buyer mixed with other non-conforming goods – the buyer may accept the goods, accept the whole of the delivery or may reject the whole of the delivery.)

Rights and actions: sections 38–54

Sections 38–48 deal with three remedies relating to the goods that are available to the seller if he remains unpaid. First, for so long as the goods may remain in his possession after sale, he has a lien, the right to retain possession until he has received payment. Then, if the buyer becomes insolvent while the goods are still in transit but have not yet been delivered to him, the seller has the right to stop delivery. Finally, the seller has the right to resell any goods over which he has exercised his lien or right of stoppage.

Section 48 (A–F) deals with a number of practical remedies available to the buyer under a consumer contract.

Sections 49–54 outline the specific remedies available in sale contracts – for damages for non-acceptance or non-delivery, or for the price, and for specific performance.

Like the previous group of sections, no comments are necessary. The provisions of the Act are self-explanatory.

Implied terms: section 55

This short section is of real importance, as it deals with the relationship between a contract, the implied responsibilities of the parties under the Act and the Unfair Contract Terms Act. It provides that terms implied by law may be varied or cancelled by the parties, and that terms implied by the Act may also be varied or cancelled by the terms of the contract.

Of course, this would be subject to the terms of the contract being compliant with the Unfair Contract Terms Act.

10

Liability Exemption Clauses

10.1 Introduction

An exemption clause is a clause that either excludes or limits liability.

Before the Unfair Contract Terms Act 1977, liability exemption clauses always needed to be very carefully drafted. Since the Act it is clear that the test of reasonableness as defined by the Act has given the courts much more freedom to deal with exemption clauses. As a result leading academic lawyers have said that exemption clauses do not need to be as carefully drafted as before. However, it is probably advisable not to take too many chances. Exemption clauses should still be carefully drafted.

> *Hotel Services Ltd* v. *Hilton International Hotels* (2000) This case concerned a contract for the supply and installation of 'Robobars' in hotel rooms. They were faulty. The contract contained a clause excluding liability for 'indirect or consequential loss'. The court decided that this clause did not exclude liability for 'direct' losses. The supplier was therefore liable for the costs of removal etc.

And even when they are properly drafted they have to be reasonable –

> *Britvic Soft Drinks Ltd* v. *Messer UK Ltd* (2002) MUK supplied carbon dioxide to BSD for use in soft drinks. The contract required the gas to comply with BS 4015, which it did. However, it was also contaminated with Benzene, which made it totally unfit for use with food products. Was a clause excluding liability under the implied terms of fitness for purpose and satisfactory quality reasonable? The court decided that it was unreasonable. BSD relied on MUK and would not expect to have to test the gas before use, and MUK could pass liability on to the actual manufacturer, which it had in fact already done.

Using Commercial Contracts: a practical guide for engineers and project managers, First Edition.
David Wright.
© 2016 John Wiley & Sons, Ltd. Published 2016 by John Wiley & Sons, Ltd.

The law's approach

The law's approach is that people have a right to compensation if they are injured or damaged by the wrongful actions of someone else. So everyone should accept responsibility if he causes loss or damage to others. Therefore, any clause that tries to reduce or avoid liability should only be allowed to succeed when it can pass the most stringent tests.

This is a very negative approach. But what it means is that the professional who can write a watertight clause that will pass the tests and then impose it on the other party will always avoid liability.

However, things have changed since the passing of the Unfair Contract Terms Act. The Act provided, for the first time, a legislative framework for liability exemption clauses that deals with the two difficult areas of monopoly/oligopoly adhesion contracts and consumer protection and then allows a degree of flexibility in areas where there is a reasonable level of commercial equality.

The result is that in some areas there may be less need for carefully drafted clauses. But in other areas, particularly relating to claims for causing wrongful damage, it is still necessary to use properly written clauses.

Liability for causing wrongful damage may be of three kinds: liability for breach of contract, for failure to comply with an obligation imposed by law (essentially by statute law) and in tort. Liability for all three is unlimited in the sense that it is limited only to the amount required to provide adequate compensation for the loss caused.

Industry's approach

The commercial approach is that permanent perfection is unachievable. However hard one tries, there is always the risk, or certainty, that human error will result, sooner or later, in poor quality services or work being done or defective equipment delivered. In most cases this will not be a problem but there is always the risk of causing a major loss/claim. Even a small incident could cause substantial losses if it had the effect of stopping a factory or plant from producing for a significant period. A minor fault in computer software could cause a string of systems to crash. All that one can say is that there is the possibility of a serious claim in almost every contract.

So the commercial approach has to be one of risk management, coupled with asset protection, where possible accepting only the risks that one can manage properly and refusing to accept potential liability if it is on a scale that could destroy the company. Liability exemption clauses are one aspect of this.

The basic position

The Unfair Contract Terms Act does not change any of the previous law concerning the validity of exemption clauses.

So the parties to a commercial contract may include clauses that protect either party against claims by the other, provided that any such clause –

- is properly included in the contract;
- is correctly drafted;
- complies with section 2(1) of the Act; and
- satisfies the 'requirement of reasonableness' as defined in the Act.

Exemption clauses have tended to come before the courts in three different types of situation –

- the classic consumer contract case, where a commercial organisation has used its superior knowledge and bargaining power to impose protective clauses in contracts with private persons severely limiting or completely excluding its liability for breach of a term of the contract;
- the standard form, or adhesion, contract, where an organisation with an actual or effective monopoly has used its market power to impose protective clauses in its supply or purchase contracts, even with other organisations; and
- where a contract has been made between parties, commercial or other-wise, who have more or less equal bargaining power.

Invariably, too, the problem will have arisen because one party to the contract will either have failed to carry out the contract properly, or caused significant loss and/or damage to the other, and is now defending himself by invoking an exemption clause to reduce or avoid liability.

10.2 Liability under statute and under the laws of tort

One of the complications of drafting a liability exemption clause is the existence of overlapping liability. To draft a clause that excludes or limits liability for a breach of contract will give not give much protection if the breach can also give rise to a claim in tort or for breach of a legal or statutory obligation. It will help therefore to have an understanding of the nature of liability in tort, and also liability for breach of statute.

Liability under statute

A number of statutes create obligations to pay compensation. There are several examples under the various Acts and regulations dealing with protection of the environment and pollution control. There is also statutory occupier's liability to lawful visitors if premises are unsafe, the Employer's Liability Act (responsibility for employee insurance against industrial injuries), the Consumer Protection Act (product liability) and the Road Traffic Acts (injuries caused by unsafe working practices on the highway), to name but a few.

Liability in tort

Liability in tort aims to compensate the victim for damage or injury that has been caused to him by another. Tort covers a wide range of separate individual torts. There is defamation, the torts of libel and slander. There is deceit (or fraud). There is trespass, which is interference with the rights of another, including rights to goods, to land, or to the person (assault, battery and false imprisonment). There is nuisance, both public and private; liability for having unsafe or dangerous areas on one's land; strict liability for damaging the land of a neighbour by allowing something harmful to escape; and so on.

But the principal tort, in the sense that it is by far the most common, is that of negligence.

This is not the place to go into all the detailed rules of the many different torts, but it may be helpful to give a brief summary of the principles of negligence.

Liability in negligence is based upon the principle of 'causal proximity'. If any person can foresee that someone else is 'in the line of fire', likely to be injured or damaged in some way if he acts negligently (or recklessly), then he owes a duty to that person to take reasonable care. If he fails to take reasonable care and as a result that other person is injured or suffers damage, then he will be liable to compensate that person for all the foreseeable losses that he has actually caused to him.

The principle is best explained by giving examples.

- When I drive a car I can foresee that if I drive negligently I may cause an accident involving damage or injury to another road user. Road users will include the drivers and passengers in other vehicles, cyclists and pedestrians. They will also include the owners of vehicles and of any goods being carried in those vehicles, and of the property alongside the road. If then I do drive negligently and I do cause an accident as a result of which other road users suffer damage or injury, I am in breach of my duty of care to those road users, who may therefore claim damages from me.
- If I supply equipment to a customer knowing that he will use that equipment or resell it to someone else who will use it, I can foresee that if the equipment is defective it may cause damage or injury to any property or person in proximity to it. If I fail to take reasonable care in manufacturing that equipment so that it is defective and causes an accident or loss of production, then I am in breach of my duty to them.
- If I give incorrect advice to a client knowing that he is depending upon me to give him advice that is correct and that he will act on the basis of my advice and may suffer loss if that advice is incorrect, then I must take

reasonable care to give advice that is correct. If I fail to take reasonable care and as a result give incorrect advice and my client then follows my advice and does suffer a loss, then I am in breach of my duty to that client.

It will of course be obvious that in the last two cases potential liability in contract will co-exist with liability in negligence.

(Liability for giving negligent advice is a comparatively recent development. It first arose in a case in 1964, *Hedley Byrne & Co Ltd* v. *Heller & Partners Ltd*, concerning negligent advice by Barclays Bank [for which the Bank avoided liability because of an exclusion clause in its contract], and now extends to many areas of commercial advice. See for example *Smith* v. *Eric S Bush* (1990) and *Harris* v. *Wyre Forest District Council* (1990), two cases referring to liability for negligent reports by surveyors.)

The extent of liability is to compensate the victim for all loss, damage or injury of any type that was reasonably foreseeable, however great the scope (and cost) of the loss, damage or injury actually caused may be. Everything is based on causal proximity and foresight. We will be liable to people who are in causal proximity to us for the damage or injury that we can (or, as a reasonable person, should) reasonably foresee as a result of our actions.

Potential liability

The result is that the commercial organisation carries potentially serious liability if it fails to carry out a contract correctly. Claims might be made –

- for breach of contract;
- in negligence;
- in a number of other types of tort;
- for breach of a legal (usually statutory) duty.

A claim under the law of contract would be for whatever amount would be sufficient to compensate the claimant for the loss of his contract. Other claims would be for whatever amount would be sufficient to compensate for the loss/injury caused.

10.3 Liability exemption clauses – general

Liability exemption clauses may try to limit liability in a number of different ways –

- by simply excluding one or more areas of liability;
- by accepting liability, but only up to a fixed limit;

- by imposing limits on the types of redress available to the injured party;
- by imposing procedural restrictions or time limits on any claim.

The attitude of the law may change according to the methods adopted. An attempt to exclude liability may well be struck down, whereas an attempt to limit liability would be held to be reasonable.

This is an area where the law can seem to be both confusing and confused.

The law will always produce a result that is completely logical, but sometimes the result in one case can appear to be based upon a totally different logic to that which is applied in another. Legislation can appear to conflict with the approach of the common law and with other legislation. The language adopted by judges can be contradictory. Different judges can take different views of the facts and of the correct interpretation of the contract. And this is the area where the law plays with words, more than any other.

The law is confusing because it is complex.

But the basic principles were (and still, in most circumstances, are) as follows.

1. The law does not accept that any party to a contract will give up his legal rights to terminate the contract or claim damages for breach of contract, or for a statutory right, or in tort unless the contract states very clearly that he has agreed to do so.
2. Therefore, any clause excluding or limiting the legal rights of any party to a contract against the other in respect of any breach of contract must be part of the terms of the contract.
3. In addition it must be both wide enough and precise enough to cover the actual situation that has arisen.
4. If it is not wide enough it will be ineffective; it will not protect at all against the loss or may only protect against one type of liability that may arise in respect of the loss but not against others.
5. If it is not precise the court can, if it wants, adopt whatever interpretation is most against the interests of the party that put the provision into the contract, again reducing or nullifying the effect of the clause.
6. Even then the clause will be ineffective unless it can be shown to be no more than is reasonable.
7. Any commercial organisation is expected to be sufficiently skilled to be able to write its contracts properly and with precision and to understand what they mean when correctly interpreted.
   That being said, however –
8. Businessmen are not expected to behave or express themselves with the same precision as lawyers.

9. There should be freedom of contract between equals, so that the parties to the contract will be free to agree whatever split of the risks and liabilities that might arise under it that they wish.

10. A clause in a commercial contract that excludes or limits the potential liability of one party can be simply part of a risk management policy avoiding double insurance.

11. It is the right of the parties to the contract to decide what terms of contract that they wish to agree.

12. When the parties are roughly equal in status then the law should not intervene more than necessary.

13. The law is only too well aware that in many specialised areas of contract, such as banking, insurance, construction and so on, the law, or more particularly the lawyer, should not interfere too much with established practices, since the experts know their business better than any lawyer.

14. In the same way the law will generally not challenge the accepted interpretation of contract conditions that are demonstrated to be in accordance with good practice within any industry.

The inclusion of the exclusion/limitation clause in the contract

In the normal commercial contract this will usually not be a problem. Where the contract is written there will be no difficulty in deciding whether or not the exclusion clause is part of the contract. It will either be part of the contract conditions (whether directly or by reference) or it will not. Even where a commercial contract is made orally, the parties will be expected to understand that written documents will form part of the contract and that they will usually contain conditions of contract, or that the parties are dealing on the basis of standard terms.

There is also no problem with a non-commercial written contract.

However, there can be difficulty where a contract is made orally but at the time the contract is made one of the parties seeks to include terms set out in writing, either in a notice or in something which is handed by one party to the other at some point. There can also be a problem where the exclusion clause is contained in a document which may or may not form part of the contract between the parties, and where one of the parties claims that there is a course of dealing between them.

Standard documents

Conditions contained in any document issued by one party to the other at the point of contract will be effective if it is reasonable to assume that the document is contractual. However, they will not be effective if the recipient would not assume the document to be contractual or to contain conditions. For example, the typical cash register/till receipt resulting from a purchase in a shop or supermarket is generally accepted as a proof of purchase document, listing the goods purchased and the prices paid for each item. But that is all it is. It is not regarded as a contractual document. It is very unlikely that any conditions printed on a till receipt would ever be accepted as part of the purchase contract between shop and customer.

Where tickets are concerned we have already referred to the case of *Chapelton* v. *Barry UDC* (see Chapter 3 for the facts). In that case the ticket given to C when he paid the charge for his deckchair contained on the reverse side a clause excluding any liability of the Council. However, the court held that the clause did not protect the Council from liability. First, it was not actually given to C until after he had entered into the contract to hire the chair by taking it from the stack. Second, the court held that it was not reasonable for C to expect the ticket to be anything more than a mere receipt for the 2p that he had paid.

In the same way a ticket issued by an automatic machine in a car park has also been held not to be the sort of document that any reasonable person would expect to contain contract conditions, so that conditions printed on the ticket were ineffective; see the case of *Thornton* v. *Shoe Lane Parking Ltd* below. This case turned on whether or not the writing on the ticket issued by an automatic machine after T had driven his car up to an entry gate formed part of the contract. The decision of the court was that the ticket was issued to T automatically, as part of the process of raising the barrier to admit him to the car park. What was written could not be disputed by him. Therefore, it was not to be expected that he should treat any writing on it as contractual, except the entry time shown on the ticket. As far as he was concerned it was merely a document that gained him entry for his car, and would then be used as a part of the payment process when he returned to collect his car later.

However, the courts have taken a different view in relation to tickets issued for more significant purposes, such as for air or rail travel.

The principles were laid down originally in a case relating to the deposit of a suitcase at a left-luggage office.

Parker v. *SE Railway* (1877) P, a conjurer, deposited a suitcase containing some of his stage props at a railway station left-luggage office, paid the fee of 2p, and was given a large numbered and dated ticket/receipt. On the front of the ticket the office opening hours were printed together with the words 'SEE BACK'. On the back of the ticket were printed several conditions, including one that limited the SER's liability to £10 per package. The suitcase was lost and P claimed £25. He had not read the conditions. Was SER entitled to assume that P would have read the conditions, and therefore accepted them? The court held that SER was entitled to assume that P could read English and would act with reasonable prudence. (The court was also clearly of the opinion that P should have realised that the ticket did contain conditions).

A similar case was –

Thompson v. *LMS* (1930) T was an elderly lady who took a railway excursion from Manchester to Darwen. She was injured when leaving the train at Darwen and claimed damages. T herself could not read, but the ticket was actually bought by her niece (as her agent) and the niece could read. On the face of the ticket was printed 'Excursion, for Conditions see back' and on the back of the ticket were words referring to the company's conditions of contract which were printed in the LMS timetables and excluded liability for claims in respect of injuries suffered during excursion journeys. The court held that in the circumstances anyone buying a ticket, and especially an excursion ticket at a mere fraction of the price of a normal ticket, would expect the railway to impose conditions of some sort, and the ticket did refer in clear terms on its face to the fact that there were conditions that would apply. Therefore, T was bound by the conditions and her claim consequently failed.

(Following the Unfair Contract Terms Act, T's claim would of course probably succeed, as it is not now possible to use contract conditions to exclude or limit liability for causing personal injury.)

The deciding factor in *Thompson*'s case was that T/her agent purchased the ticket from a ticket clerk, and therefore had the opportunity to refuse to accept the ticket when she saw that it contained a reference to conditions.

Notices

Chapelton v. *Barry UDC* has also already given us an example of a contract based upon the contents of a notice. Where the notice was displayed so that it is visible to any person wishing to make a contract, the law would expect the reasonable person to read that notice before making that contract and therefore to be bound by its terms. But what if the notice is not displayed where it can be seen at or before the time that the contract is made? The classic example is the following case –

Olley v. *Marlborough Court Hotel* (1949) Mr and Mrs Olley took a room at the MCH. When they arrived they went to the reception desk. They could see behind the reception desk a notice stating the prices of rooms. After they had booked into the hotel, they then found a further notice inside a cupboard in their room setting out a number of conditions, including one that stated that MCH took no responsibility for anything stolen from a guest's room. Later, due to the negligence of the hotel in leaving the reception desk unattended, a thief walked into the hotel, took the keys to Mr and Mrs Olley's room from behind the desk and stole a quantity of clothing, jewellery and a valuable fur coat. The court decided that the contract was made when the Olleys booked themselves into the hotel at the reception desk. A notice in the room, which they could not have seen until after they had made the contract, could not form part of the contract.

Dealings within the trade

Where the parties are both in the trade it is to be expected that even an informal contract will be based upon the usual contractual framework within the trade, especially relating to areas of risk management; see *British Crane Hire Corporation Ltd* v. *Ipswich Plant Hire Ltd* (1975) concerning an informal contract to 'borrow' a mobile crane. Since both parties were in the plant hire business it was to be expected that the standard hire conditions would apply.

Inconsistent use and courses of dealing

Here the situation will depend upon whether a 'course of dealing' has been established between the parties –

> *Spurling* v. *Bradshaw* (1956) B regularly stored goods at a warehouse owned by S. Every time he deposited goods he was sent, a few days later, a 'landing account', listing the goods that had been put into store and the storage charges, and including on the back a small-print set of conditions of contract, which included what was known at the time as the 'London lighterage clause'. This was a comprehensive clause used within the warehousing industry which excluded the liability of the owner of the warehouse for any damage to goods stored within his premises. B deposited barrels with S; they were damaged during storage and the contents lost. B claimed damages and S responded that the exclusion clause protected him from the claim. The court had no difficulty in finding that the exclusion clause applied. This was a commercial contract, and there was a regular course of trading. Even though B had never bothered to read the small print, he must have known that it contained conditions of contract and that S always invoked them. They therefore applied. (The court also noted that there was nothing exceptional in conditions like this in commercial contracts. In fact, all that they provided was that it was the responsibility of B to make sure that he had proper insurance cover for his goods while in store.)

By contrast –

> *McCutcheon* v. *David McBrayne Ltd* (1964) From time to time, M, a crofter, shipped cattle, goods or his car from the island of Islay to the mainland by DM's ferries. DM's procedure was that anyone using the ferry should complete a booking form giving details of whatever was to be shipped. The booking form referred to conditions of contract displayed at the harbour offices and contained a clause excluding DM's liability for any items being carried. Then the owner was to be given a receipt which he was required to sign, together with a 'Risk Note', in effect a written contract also containing the exclusion clause, which he was also asked to sign and which again included DM's conditions of contract. One might say that this procedure was fairly watertight.
>
> M asked his brother-in-law, McSporran, to take his car to the mainland. When McSporran boarded the ferry he did not complete a booking form, nor was he given a Risk Note. He was only given a receipt.
>
> It then emerged that DM's policy was applied rather haphazardly; on some occasions McSporran had completed a booking form and been given a Risk Note, but on several others he had not.

Through the negligence of DM the ferry hit rocks and sank, taking the car with it. M sued for the cost of the car and the court held that there was no course of dealing between the parties, and that in effect, because no Risk Note had been signed, there was only an oral contract between the parties.

Therefore, DM could only invoke its exclusion clause if it could show that M (or McSporran) had been made aware of the exclusion clause and that DM intended it to apply. This DM totally failed to do. DM's representative at the booking office had not mentioned any conditions. McSporran had never read the booking form or the Risk Note, was unaware that DM's conditions were displayed in the harbour offices and had never looked at them (and probably would not have understood them if he had read them). DM was therefore liable.

10.4 The interpretation of exclusion and limitation clauses

The common law rule is that liability exemption clauses are to be strictly interpreted, usually on what is termed a *contra proferentem* basis. *Contra proferentem* simply means 'against (the interests of) the person who put the clause into the contract'. So if there is any mistake in the wording, for example, if the words are not totally comprehensive or precise, or if they are ambiguous, then the mistake will be resolved against the interests of that person, who of course will usually be the person who wants to rely on the clause.

There are only two situations where exemption clauses may be interpreted less strictly –

- where the clause is one that accepts liability for the loss caused but then seeks merely to limit the extent of liability for that loss;
- where the conditions of the contract which include the exemption clause are model conditions, accepted within the industry concerned as being a fair and reasonable basis for contracts within that industry (and model conditions tend to be fairly well drafted anyway). But see the case of *Walker* v. *Boyle* referred to below.

To repeat, the consequence for the commercial organisation is simply that if it intends to base its contracts on conditions which it has drafted and which include an exemption clause, then that clause must be drafted correctly.

10.5 Exemption clauses and liability for breach of contract

What breaches may be covered by a liability exemption clause?

There are two kinds of breach/risk. There is the failure to carry out the contract properly, by supplying something late or supplying something that does not comply with the specification, or does not meet its guarantees, causing financial loss. There is also the situation where work is carried out wrongly or defective equipment is supplied and causes physical damage or injury.

10.6 'Fundamental breach'

In the years following the Second World War there was considerable debate in the courts about whether law should permit an exemption clause to exclude liability for a total failure to carry out basic contract obligations. This approach was advocated in particular by Lord Denning. He tried to introduce a new rule into the law of contract that would strike out any clause that sought to exempt liability for 'fundamental breach', or breach of a 'fundamental term' of the contract, that is, a failure to carry out any significant term of the contract.

The doctrine achieved considerable influence during the 1950s and 1960s. But it was pointed out that we cannot always guarantee to achieve success. In the field of medical research, for example, we can try to develop a cure for, say, the Ebola virus (and we know that sooner or later we will probably find it), but we can never guarantee that it is this piece of research that will find it. So the doctrine was struck down in the end by the House of Lords in a case in 1966, *Suisse Atlantique Societe D'Armement Maritime SA* v. *N V Rotterdamsche Kolen*. The court held that it was permissible for a clause in a contract to exempt one of the parties from liability even for a total failure to comply with the terms of the contract, provided that the wording of the clause was adequate to do so. (And to avoid liability for a complete failure to comply with the terms of the contract, the wording of the contract clause would need to be very adequate indeed.)

In commercial contracts one can get situations where it is perfectly proper for one party to wish to exempt itself from failure. For instance, it is normal practice to include a *force majeure* clause to protect against delay in delivery or completion due to circumstances that cannot be controlled.

The general approach to exemption clauses is that every clause dealing with contractual responsibility will be judged solely on its merits. There is any number of examples of how the courts deal with this area. It is sufficient to give just a few –

Beck & Co v. *Szymanowski (K) & Co* (1924) This concerned a contract for the supply of '200 yard reels' of sewing cotton. A clause in the contract provided that 'the goods delivered shall be deemed to be in all respects in accordance with the contract and the buyer shall be bound to accept and pay for the same accordingly unless the sellers shall within 14 days after the arrival of the goods at their destination receive from the buyers notice …' that there was something wrong with the goods. The reels actually delivered held significantly less than 200 yards of cotton, but B did not discover this until sometime after delivery. B claimed for the short delivery. The court ruled that '200 yard reels' did not mean that each reel should contain precisely 200 yards of cotton, but did mean that overall B should receive close to 200 yards per reel. The court then held that the reference in the exemption clause to 'the goods delivered' could only apply to goods that were delivered and complied with the contract. Because the reels delivered contained significantly less than 200 yards of sewing cotton they did not comply, therefore the clause did not apply and B's claim did not need to be made within the 14-day period stated in the clause.

Andrews v. *Singer* (1934) A contract between a car manufacturer and a dealer contained a clause that excluded the manufacturer's liability for breach of 'implied liabilities, warranties and conditions' of the contract. The contract was for the supply (and resale) of 'new Singer cars'. S delivered a car that was not new. The court held that a clause excluding liability for implied conditions could not protect S against his liability for breach of an express condition.

Wallis, *Son & Wells* v. *Pratt & Haynes* (1910) A contract for the supply of seed to a farmer contained a clause excluding liability for 'breach of warranty'. The contract was for the supply of seeds for 'common English sainfoin'. What was actually supplied was seed for 'giant sainfoin' (which is very different in agricultural terms). The court held that the clause only protected against a breach of a warranty, not a breach of a condition. So it did not protect the seedsman against a breach of the implied condition under section 13 of the Sale of Goods Act that the goods must comply with the contract description.

Thornton v. *Shoe Lane Parking Ltd* (1971) T, a leading classical trumpet player, who was due to play in a BBC recital, parked his car in an automatic car park. He had not used it before. A notice at the entrance stated the charges for parking, the opening hours and so on, and then warned that cars were 'parked at owner's risk'. As he had entered the car park, an automated machine issued him with a timed ticket. T glanced at the ticket to check the time printed on it, but did not read any further. In fact, the ticket contained a statement that parking in the car park was subject to the SLP conditions of contract, which were actually displayed on a noticeboard at the other end of the car park from the entrance used by T, and had not been seen by him. They excluded the liability of SLP for any injury suffered by a user of the car park. On returning to the car park after the recital, T was injured in an accident through the negligence of SLP. T sued for damages and SLP claimed that the exclusion clause in their conditions protected them from his claim. The court held that the only conditions of the contract were those set out on the notice that could be seen by T when he entered the car park, and that the clause stating that cars were to be parked at the owner's risk excluded the liability of SLP for damage to vehicles, but was not wide enough to cover the risk of injury to the drivers of those vehicles or their passengers.

10.7 Exemption clauses and liability in tort, including negligence

Interpretation

To repeat what we have already said, a breach of contract may create several other kinds of liability. Quite apart from the possibility that the breach might be a crime, which of course does not concern us here, it could also be a breach of an obligation under statute, or create liability in one or more torts. The result of this is that the injured party may have several different potential remedies that he can use in order to claim compensation from the defendant. If this is the case, because the injured party can bring several different claims against the defendant each of which would give him compensation for his loss or injury, the law is that an exemption clause will only be successful if its wording is both wide enough and precise enough to cover all the different heads of claim that the injured party has available.

So, for a clause to protect a defendant from liability, it must, in practice, take one of two possible approaches.

The first approach is to exempt the defendant from liability for breach of contract, then in tort generally, then specifically in negligence, and then also 'in law'. This approach is reasonably certain to be successful, provided that the clause is correctly written. See the rules set out in *Canada Steamship Lines Ltd* v. *The King* below.

The second is to use other language that is wide enough to cover negligence as well as all other possible heads of claim. Exemption clauses of this type that have been accepted by the court include phrases such as 'at the sole risk of xxx', 'at their own risk … ', 'not under any responsibility for any injury, loss or damage howsoever caused … ', or 'which may arise from or be in any way connected with any act or omission of any person … ' However, this type of clause is inherently less certain to provide exemption, simply because it is always open to interpretation in the context of the rest of the contract (see below).

Clearly this is logical but rather artificial. The policy was of course originally developed by judges looking for maximum power to strike down what they felt to be unfair exemption clauses. However, it is now enshrined in the law, even though the Unfair Contract Terms Act gives the court a much better approach to use. Again, the consequence for the commercial organisation is that clauses do need to be precisely drafted.

The way that a court should interpret an exemption clause that deals with liability in tort was defined in a case concerning damage to goods stored in a warehouse caused by the negligence of a contractor carrying out work for the warehouse owner: *Canada Steamship Lines Ltd* v. *The King* (1952). The House of Lords (as the Privy Council) set out the following rules.

1. For an exemption clause to be allowed to succeed it must exclude or limit liability in clear language.
2. If the clause clearly exempts the defendant from liability for negligence, the clause will then be effective.

3. But if there is no express reference to negligence the clause will not be effective unless the words are wide enough in their ordinary meaning to cover loss and damage caused by negligence.
4. If the words are wide enough to cover negligence, but in the circumstances of the contract it would have been reasonably possible for loss or damage to have been caused in another way (in this case for example by the deliberate acts of the warehouse owner), then the court will automatically assume that the exemption clause is not intended to exclude or limit the defendant's liability for negligence as well as those other possible causes.
5. This is unless, of course, the clause exempts the defendant from liability both from negligence and from those other possible causes as well.

Again, the best way of explaining how these principles work in practice is by examples –

In the case of *Olley* v. *Marlborough Court Hotel* (1949) (see above for the facts) the wording of the notice was 'The proprietors will not hold themselves responsible for articles lost or stolen unless handed to the manageress for safe custody'. Although it was not necessary for the decision, the court did comment on the clause: 'The defendant is not exempted from liability for negligence unless adequate words are used … If (MCH) wants not to be liable it must show that … the words are so clear that they must be understood by the parties in the circumstances as absolving the defendants from the results of their own negligence' (which they were not).

White v. *Warwick (John) & Co Ltd* (1953) This case concerned a contract to hire and maintain a 'tradesman's bicycle', a bicycle adapted to carry heavy loads for local delivery. The exemption clause was 'Nothing in this Agreement shall render the owners liable for any personal injuries caused to … ' any person using the bicycle. Due to poor maintenance the saddle suddenly tipped while the bicycle was being ridden, causing injury to the rider. The court held that the wording did not protect Warwick against a claim for negligence.

Hollier v. *Rambler Motors (AMC) Ltd* (1972) The case concerned a clause excluding the liability of a garage for damage caused to any vehicle by fire. That clause read simply '[t]he company is not responsible for damage caused by fire to customers' cars on the premises'. The court held that the clause was not adequate to protect against a claim for negligence:

It is well settled that a clause excluding liability for negligence should make its meaning plain on its face to an ordinarily literate and sensible person. The easiest way of doing that, of course, is to state expressly that the garage, tradesmen or merchant, as the case may be, will not be responsible for any damage caused by his own negligence …

10.8 Liability limitation clauses

A liability exemption clause (as opposed to a liability exclusion clause) can set out to achieve its objective in more than one way. We have already seen the approach to a clause that seeks to exclude claims by imposing procedural rules (see *Beck*'s case above) rather than by excluding liability. (The court interpreted the clause against the interests of the defendant.)

The approach to clauses which limit rather than exclude liability is much less strict. An early example of this is the following –

> *Joseph Travers & Sons Ltd* v. *Cooper* (1915) The case concerned goods that had been damaged by C while being shipped by his barge. Contracts of this kind, under common law, required the carrier to accept responsibility for negligence, unless excluded by the contract. The contract contained a clause excluding C's liability for 'any damage to goods however caused which can be covered by insurance'. The clause clearly set out to exempt C from liability, but did not mention negligence, nor did it contain any words of the kind referred to in *Canada Steamship Lines Ltd* v. *The King*. Nevertheless, the court found that the clause was perfectly acceptable. It accepted that C was liable for losses caused in general terms, but simply provided that where protection could be provided by insurance cover then JT should provide that cover, and accept responsibility for the cost of any loss.

A much more recent example of the same approach is the following case, which is of great importance as it sets out the general approach of the courts following the Unfair Contract Terms Act –

> *Ailsa Craig Fishing Co Ltd* v. *Malvern Fishing Co Ltd* (1983) The case resulted from the sinking of two fishing boats, owned by ACF and MF respectively, in the harbour at Aberdeen. Securicor had entered into a contract with the owners association to provide security and oversight of their vessels while moored in the harbour. Due to the negligence of S, the two vessels became trapped underneath a jetty on a rising tide. Before they could be released, they were swamped and sank. In a series of actions ACF and MF sued each other and then sued S.
>
> S's contract contained two exemption clauses. The first set out to exclude all S's liability. This was rejected by the court. The other clause limited S's liability, 'for any liability whether under the express or implied terms of this contract or at common law, or in any other way …' to £1000. The court accepted the validity of this liability limitation clause, even though its wording was very simplistic (for instance, it made no reference at all to liability 'both in tort generally and in negligence'):
>
>> The principles of contra proferentem interpretation of exemption clauses against the interest of the defendant should not be applied so harshly against [a] clause that merely seeks to limit liability. Such a clause should of course be read against the interests of the defendant and must be

clearly written, but [it] should not necessarily be judged by the especially exacting standards which are applied to exclusion or indemnity clauses. The reason for imposing rigorous standards to an exclusion clause is the inherent improbability that any party to a contract would totally release the other from liability that would otherwise fall on him. But there is no such high degree of improbability that he would agree to a limitation of liability, especially when … the potential losses that might be caused by the negligence of [S] or its servants are so great in proportion to the sums that could reasonably be charged for the services contracted for. It is enough in the present case that the clause must be clear and unambiguous.

And of course the decision was again influenced by the fact that the owners would be expected to insure their boats.

10.9 Statutory control of liability exemption clauses and the Unfair Contract Terms Act 1977

The Act has been very successful in achieving its objectives.

It followed on from recommendations made by the Law Reform Commission in 1975 concerning liability exemption clauses in commercial contracts in general, and in adhesion contracts in particular, and consumer protection. The Commission wanted more to be done to control the 'unfair' use of liability exemption clauses by commercial organisations with excessive bargaining power.

The Act has done this by creating a simple general rule that in any contract between a business and a business or consumer based on 'standard terms of business', liability exemption clauses will be effective only to the extent that they are reasonable.

Its title is misleading. It is not limited just to terms in contracts; it also applies to various non-contractual statements, such as notices. Also it does not deal with unfair terms in general, only with liability exemption clauses.

It does not change the law relating to the basis of liability, or the law on preliminary issues such as whether the clause is included in the contract or its proper interpretation. It is concerned solely with the question of whether or not any party to the contract will be allowed to use a clause in that contract to limit or exempt himself from liability. It controls exemption clauses relating to 'business liability', not clauses in contracts between private individuals.

The Act itself is confusing to read, because it covers a large number of different possible situations, and tends to repeat itself. But it lays down a number of simple principles, and then expands the detail surrounding those principles to make it extremely difficult to avoid them. It deals specifically with four types of clause:

- clauses that seek to avoid liability for personal injury or damage etc. caused through negligence;
- clauses that seek to avoid liability for breach of terms imposed by common law or statute;
- terms in consumer contracts that avoid liability; and
- standard terms that avoid liability.

The definitions used in the Act

'Negligence' means negligence under the law of tort, plus the breach of any express or implied obligation under a contract or under law generally to take/exercise reasonable care or skill, plus occupier's responsibility to ensure safe premises (section 1(1)).

A 'business' means any person or organisation that engages in any way in commercial operation. It does not have to be a commercial profit-making business. It includes, for example, the operations of any charity, university, professional organisation, or national or local government body. Then 'business liability' is defined as the liability for breach of obligations or duties arising from things done in the course of a business, or from the occupation of any premises used for business purposes (section 1(3)).

A 'consumer' is a person (or company) who contracts for or buys goods, of a kind that is used for private use or consumption, and does not do so in the course of a business (section 12).

'Consumer use' is the use of goods otherwise than exclusively for a business (section 5).

The 'requirement of reasonableness' means that any person wishing to rely on a liability exemption clause must be able to demonstrate that it satisfies the standard laid down by the Act. The clause must be fair and reasonable having regard to all the circumstances that the parties should or did have in mind at the time when the contract was made (section 11). The section then refers to Schedule 2 to the Act, which lists a number of relevant factors.

1. The strength of the bargaining positions of the two parties, and whether the customer could have purchased his requirements elsewhere, perhaps on a different basis.

   A limitation of liability imposed in an unequal bargaining situation, especially when the customer had nowhere else to go (such as an adhesion contract in particular), is very unlikely to be reasonable.

2. Did the customer receive any inducement to agree to the clause, or could he have entered into a similar contract with another without having to accept a similar clause?

   If the limitation of liability was purchased by a price reduction, or if the customer could have avoided having to enter into the clause by buying elsewhere, but chose not to do so, then that clause is very likely to be reasonable.

3. Whether the customer knew or should reasonably have known of the existence and extent of the clause, bearing in mind the custom of the trade, and any previous dealings between the parties.

This raises the question of commercial understanding. The greater the understanding of the customer of the true nature of the liability exclusion or limitation clause, the more likely it is that he accepted it willingly, and therefore that it is reasonable in the circumstances (although of course this does not necessarily follow if there is a serious inequality of bargaining power – see factor 1 above).

4. Where the clause excludes or limits liability for breach of a contract condition, whether it was reasonable at the time of the contract to expect that compliance with that condition would be practicable.

This is what might be called achievement/failure risk. The more difficult it is to comply with a contract condition, and therefore the higher the risk of failure, the more reasonable it is for the customer to expect/accept that the supplier will not be prepared to take that risk.

5. Whether the goods were manufactured, processed or adapted to the special order of the customer.

This relates to the question of the customer's requirements and innovation risk. The greater the degree of innovation in any contract the greater the degree of risk taken by the supplier, and therefore the more reasonable it might be for the supplier to seek some limit to his liability for that risk.

6. Added by section 11(4) of the Act: in the case of clauses limiting, as opposed to excluding, liability, the availability of insurance cover and the resources that the parties might have available to deal with the risk.

In other words, the greater the ability of the defendant to insure his liability and the greater his financial resources (when compared to those of the customer), the less likely it will be that any limitation of liability clause will be reasonable, unless the limit of liability fixed by the clause is reasonably adequate in the circumstances of the case.

The principles

Liability for negligence

No one may exclude or limit his liability for death or personal injury caused by his negligence. No one may exclude or limit his liability for any other loss or damage caused by his negligence except to the extent that it is reasonable.

Remember that the definition is much wider than simply the tort of negligence. It includes all contractual obligations, including those that are implied by law, such as the duties of warehousemen and carriers already referred to, and the duties of the occupiers of premises to make them reasonably safe for visitors and to give warning of hazards, and so on.

Title to goods

The implied condition under section 12 of the Sale of Goods Act that the seller has the right to pass title in the goods to the buyer cannot be excluded or limited.

> In other words, if the seller does not have unchallenged ownership of the goods or cannot pass title and quiet possession to the buyer, then this must be clearly stated in the contract, so that an express term of the contract will over-rule the implied term.

Contracts

International contracts and auction sales

The Act does not apply to auction sales. Nor does it apply to international contracts. But it will apply to any UK contract, even if the parties have tried to avoid the Act by making it subject to a different system of law.

Contractual liability in consumer contracts

(These principles are now subject to the Consumer Rights Act 2015.)

In a consumer contract the business cannot use any term of the contract –

- to exclude or restrict his liability for a breach of contract, or
- not to carry out any part of the contract at all, or
- to carry out the contract in a way that was substantially different from what the other party had a reasonable right to expect,
- except to the extent that the term satisfies the requirement of reasonableness.

And

> In a consumer contract, any clause that requires the consumer to indemnify someone else against their liability for breach or negligence must satisfy the requirement of reasonableness.

And

> The consumer's right to claim against a manufacturer or distributor of goods for breach of the 'manufacturer's warranty' in respect of the goods cannot be defeated by any term in the contracts of sale between the manufacturer/distributor and retailer, or the retailer and the consumer (though they may prevent claims between businesses within the supply chain).

And

> Any attempt in a consumer contract to exclude liability for breach of the implied conditions under the Sale of Goods Act etc. relating to conformity

of the goods with description/sample, or their quality or fitness for purpose, will fail.

And

These principles apply to all consumer contracts, whether of sale, hire purchase or credit sale, and so on.

This has put an end to the standard boilerplate clauses in consumer contracts that avoided seller's liability.

The effect on consumer contracts in general was to eliminate the exclusion clause as a practical possibility, except in extremely unusual circumstances. Virtually the same position has emerged in relation to liability limitation clauses. In fact, the Act led immediately to the wholesale abandoning of such clauses in consumer contracts, both in 'one-off' contracts and contracts of adhesion, so that there are very few case examples.

One of the few, *Walker* v. *Boyle* (1982), was concerned with the sale of a house under the Law Society's National Conditions of Sale, used generally as the standard conditions within the conveyancing industry. Condition 17 stated that there was no guarantee by the vendor that the replies that he gives to preliminary enquiries by the purchaser are accurate. The vendor responded to a preliminary enquiry as to whether or not there were any disputes concerning the property that there were none. In fact, there was an ongoing serious boundary dispute, of which she was aware, or at the very least should have been aware.

The court held that the fact that the conditions were produced by the Law Society did not prevent them from being 'standard terms of business' for a solicitor. The court then held that the condition was unreasonable, so that the vendor was liable.

Contractual liability in business/commercial contracts

Where one of the parties to a business contract deals on the other party's 'written standard terms of business', any contract term that seeks to exclude or limit the liability of the other party for any breach of or failure to perform the terms of the contract properly will be ineffective unless it satisfies the requirement of reasonableness.

This principle is aimed particularly at monopoly/oligopoly adhesion contracts, but it may apply to other contracts as well.

It raises three issues –

- When will a contract be made on the basis of one party's standard terms of business?
- What criteria will a court apply in deciding whether any term is reasonable?
- How will those criteria be applied in practice?

10.10 Issue 1 – standard terms of business

The vast majority of business contracts are made on the basis of pre-existing terms put forward by one of the parties. Often those terms are then simply accepted by the other party without any real discussion. In some areas, however, there may be considerable negotiation and many changes to the conditions before the final contract is agreed.

A wide range of such terms may be used, ranging from the (very) small-print standard conditions of purchase/sale/hire etc. that appear on the backs of standard business paperwork, through sets of contract conditions, often produced by trade associations, that are used as standard terms for buying and selling everything from hardware to holidays, through the industry-wide conditions used in warehousing, transport, shipping and equipment hire, up to the 'model' sets of conditions drawn up within the various engineering industries as a deliberate exercise in producing contract terms that are fair to both sides.

Obviously in many cases there will be very little doubt whether or not a contract will be based upon the standard written terms of business of one party or the other. Where goods have been sold on the seller's 'standard conditions of sale', for instance, there could be no serious dispute. There will be much more doubt if a contract has been made on the basis of industry-standard conditions that both parties are equally happy to accept.

The courts have solved the question by simply refusing to discuss the matter. They have taken the simple approach that, whatever the conditions may be, when one of the parties proposes them for the contract he will adopt those conditions as his own written terms of business. The question of where the conditions originated, and how fair they are, will be considered as a part of the answers to issues 2 and 3. Unless it can be shown that the contract was drafted as a wholly separate exercise, the conditions of the contract will automatically be treated as the standard terms of whichever of the parties proposed them.

Walker v. *Boyle* referred to above is an example. Even though the National Conditions of Sale were the conveyancing industry standard, automatically adopted by the seller and probably by both sides in almost any contract to sell real property, the court had no hesitation in treating them as the seller's standard written terms of business, and then concentrating on whether or not condition 17 was actually reasonable. What is even more interesting is that even in the circumstances, where the conditions were used widely, where both parties were professionally advised and bargaining power was approximately equal, the court was prepared to hold that the clause was unreasonable.

In other words, the court was fully prepared to adopt the view that even though a set of conditions of contract is standard within a particular industry and has been written with the intention of being reasonably fair, if an exemption clause fails to achieve fairness it will be unreasonable within the meaning of the Act.

10.11 Issue 2 – reasonableness criteria

The courts have simply adopted the criteria laid down in the Act, including the tests set out in Schedule 2, as the starting point for deciding whether any clause is reasonable. For instance, in the *Stag Line* case below the court referred to paragraph (d) of Schedule 2 as one of the factors in deciding that clause 8(4) was unreasonable.

10.12 Issue 3 – practical application

A liability exemption clause may be simple, comprising a single statement of exclusion or limitation of liability, or it might be complex, containing any number of separate statements. Clearly, in the case of a simple liability exemption clause, if the exclusion or limitation is unreasonable the whole clause must fail.

Where there are several clauses, or a complex clause, the courts have looked at the contract as a whole, including the overall risk management and insurance implications.

Risk management and insurance –

Photo Productions Ltd v. *Securicor Transport Ltd* (1980) This case arose out of events that took place before the Act, but it did not reach the courts, and in particular the House of Lords, until after the Act had come into force. Therefore, when the House of Lords gave its decision it was in the position of applying the law previous to the Act but being able to look forward to the position that would apply in future. The decision in fact turned upon the validity and application of an exclusion clause, which excluded all liability of Securicor for any damage to any premises being guarded by the company however caused. The facts were that an employee of Securicor, who was providing a guarding service for a factory, deliberately started a fire which destroyed the factory. The court held that the wording of the clause was watertight and therefore it applied to protect Securicor from liability. However, in an *obiter dictum*, Lord Wilberforce, who gave the leading judgment, commented upon the position that would apply under the Act. Securicor's justification for the clause was that it guarded industrial property worth several billion pounds, far more than the amount of any insurance cover that it could buy. So Securicor excluded liability for property damage in its contracts. It was clear that it was the responsibility of the owners of the property being guarded to obtain adequate insurance cover for themselves. (And in fact the case was really between the insurers of the two sides.) Lord Wilberforce said commercial companies manage risk, and in that context an exclusion clause may be seen simply as a risk/insurance transfer clause. Particularly as the sums charged by Securicor for the service were low, it was perfectly reasonable for Securicor to take the approach that it had. Any commercial purchaser of the service could read the clause, understand the position taken by Securicor and buy appropriate insurance cover. So in a situation where both sides could insure (and in fact both sides had insured) it was completely reasonable for Securicor to use the clause, not as a way of avoiding liability, but as a method for managing the risk that it took in the normal way of business.

Then there was another case where the availability of insurance was an important factor, which contrasted with *Photo Productions Ltd* v. *Securicor Transport Ltd*. The case concerned not an exclusion clause, but a clause limiting liability, and resulted from the poll tax –

St Albans City & District Council v. *International Computers Ltd* (1996) The Council purchased software from ICL to calculate and print out its poll tax demands. Due to the fact that the software was delivered late there was no chance to conduct preliminary trials and the software was used in December 1989 to produce the Council's formal demands for payment of the tax. Due to an error in the software, the number of people living within the area was overstated by some 3000 people. The result was that the individual claims for poll tax were set too low, so that the amount of tax actually paid was some £484,000 lower than it should have been. In addition, again as a result of the software error, the Council overpaid its contribution to the county council. The court held that the computer disk containing the software was 'goods' and that therefore the contract was covered by section 14 of the Sale of Goods Act. Therefore the implied conditions of satisfactory quality and fitness for purpose would be taken into consideration. The contract was actually governed by ICL's standard terms of sale, and the applicable clause limited ICL's liability to the lower of £100,000 or the contract price. ICL was clearly responsible for causing the loss to the Council. The court decided that under the Act the clause could only apply to the extent that it was reasonable. The Council could not take out insurance cover against the risk of sending out incorrect tax demands. Cover of that sort was simply not available. ICL on the other hand could and did insure itself against the risks arising from the supply of faulty software to its customers. In fact, ICL had considerably more insurance cover than the amount of damage claimable by the Council. The Court held that in circumstances where the Council could not insure, but ICL could insure and in fact did have very adequate insurance cover, the clause was unreasonable.

Then there is another case relating to insurance cover, this time in respect of the supply of the wrong goods. The case arose from a situation that occurred in 1973, but the principles are the same as under the Act.

George Mitchell (Chesterhall) Ltd v. *Finney Lock Seeds Ltd* (1983) GM was a farmer who purchased a quantity of seed for winter cabbages from FLS at a price of £192. What FLS actually supplied was seed for autumn cabbages. As a result the crop failed totally, causing a loss of £61,000 to GM. It was quite clear that the wrong seed had been supplied and that FLS was responsible. The contract included a standard 'seed trade clause' under which the liability of the supplier was limited to replacement of the seed or refund of the price. The court held that the contract was governed by section 4 of the Supply of Goods (Implied Terms) Act 1973, which provided that in these circumstances the exclusion clause would not be enforceable except to the extent that it was fair and reasonable to allow reliance on it. The evidence given to the court was that although it was normal practice for seedsmen to include the clause, it was generally recognised within the industry that it was impossible for farmers to insure against the risk of planting the wrong seed, whereas it was normal for seedsmen to insure against the risk of supplying the wrong seed. Therefore the general practice within the industry was for claims of this type to be settled by agreement for payment of compensation well in excess of what was provided by the clause. The court therefore held that it would not be fair to allow FLS to rely on the exclusion clause.

Turning now to 'model' conditions of contract. A case that contrasts with *Walker* v. *Boyle* referred to above is –

British Fermentation Products Ltd v. *Compair Reavell Ltd* (1999) CR supplied an air compressor to BFP. It failed to meet its guaranteed output. CR carried out further work, but could not remedy the shortfall. The contract was based on the model electrical/mechanical engineering industry conditions, which contained a watertight exclusion clause. In addition it provided that if the compressor failed to meet its guaranteed performance, BFP could either reject the compressor or purchase a replacement elsewhere at CR's cost. BFP refused to take either of these remedies. It terminated the contract for breach, but kept the compressor and then claimed damages of £1.3 million, approximately four times the contract price. It then claimed that the exclusion clause was unreasonable. The court held that the clause was a perfectly sensible business arrangement. It offered three solutions to a foreseeable problem, to reject, replace or retain the equipment. It had also been agreed to by BFP. The clause was therefore reasonable and BFP's claim was rejected.

And in another case where again there was a reasonable degree of equality of bargaining power, *Watford Electronics* v. *Sanderson CFL Ltd* (2001), the court said, 'Where businessmen representing good companies negotiate an agreement they are the best judges of the commercial fairness of the agreement. Unless there is clear evidence of taking unfair advantage the court should not interfere ... '

A complex clause in 'in-house' conditions, limiting liability by procedure –

Stag Line Ltd v. *Tyne Ship Repair Group Ltd* (1984) The case concerned a repair to the tailshaft (a tube through which runs the propeller shaft) of a ship. TSG replaced the lining of the shaft, but used incorrect material. The correct material would have lasted for four years. Three months after repair, while the ship was crossing Lake Erie, the lining overheated severely due to the material and also due to excessive silt in the cooling water. The overheating made it necessary to stop the engines, which brought the ship to a standstill for a couple of hours. The ship was then able to proceed, but it was clear that that further repairs were necessary. The ship completed its voyage to Europe and was then repaired in Rotterdam. SL then sued TSG for the cost of the repairs.

The relevant clause was clause 8 of TSG's standard conditions. It provided that TSG's entire liability was to repair defects and then limited TSG's liability to the cost of physical damage caused by its negligence, excluding economic loss. SL had not suffered any serious economic loss, because the interruption to the ship's voyage had been short and the repairs in Rotterdam took only a short time. Nor had the ship suffered any permanent damage. However, the court did decide that TSG were negligent and were also in breach of the terms of the contract in using the wrong material for the original repair.

Finally, clause 8 laid down the procedure to be followed by SL in the event of a defect in the work. SL was required to notify TSG with particulars of the defect. Then TSG would either carry out repairs at its yard on the Tyne or get the repairs carried out at some other yard to be selected by TSG, and SL should allow TSG to do this. The court found that this procedure was not reasonable. This was because its apportionment of the risk (or cost) of breakdown was entirely capricious, depending upon where the ship might be at that time and whether it was economic or not to return the ship to the Tyne for repair. As a result TSG had to reimburse SL the cost of repair.

In the cases of *Interfoto Picture Library Ltd* v. *Stiletto Visual Programmes Ltd* (1989) concerning a clause imposing a significant charge for delay in returning photographs, and *AEG (UK) Ltd* v. *Logic Resources Ltd* (1996) concerning a clause requiring the buyer to meet the cost of returning defective goods to the seller for repair (from the Middle East), both clauses were also found to be unreasonable because in practice they were both too onerous.

The courts have resolutely refused to discuss the principles underlying the decisions that they have made concerning the reasonableness or otherwise of various clauses. The *Stag Line* case itself is a good example of this. The court discussed the facts in reasonable detail but then ruled simply that the procedure was not reasonable. While this may be a completely accurate statement of the reason for the court's decision it gives us no explanation as to why the court decided that that reason was adequate grounds for the decision. All we have is a series of decisions in one direction or the other with very little discussion of underlying principles.

While these cases are helpful they do not give any precise guidance as to how the commercial organisation should proceed. Each case has to be considered on its own merits, and the law reports do not really tell us what arguments are accepted by the court. However, there are some reasonably clear principles which should be borne in mind by any commercial organisation when considering what liability exemption or limitation clause to include in its contracts, and how to manage any problems that may arise.

1. It is for the defendant to prove that the clause is reasonable – section 11(5) of the Act. Could we produce reasonable evidence to do this?
2. The availability of insurance cover to both sides is of importance when considering clauses to exclude or limit liability for causing damage or

loss. If it is not available to both sides it is unlikely that a clause will be reasonable – *St Albans City & District Council* v. *International Computers Ltd* and *George Mitchell (Chesterhall) Ltd* v. *Finney Lock Seeds Ltd*.

3. If adequate insurance cover is available a clause is more likely to be reasonable – *Photo Productions Ltd* v. *Securicor Transport Ltd*.

4. Liability for causing death and personal injury cannot be excluded or limited – section 2(1) of the Act.

5. A clause in 'in-house' standard conditions is much less likely to be reasonable, unless it can be justified by clear evidence – *Photo Productions Ltd* v. *Securicor Transport Ltd* and *Stag Line Ltd* v. *Tyne Ship Repair Group Ltd*.

6. A clause may be unreasonable if it tries to impose onerous charges on the other side – *Interfoto Picture Library Ltd* v. *Stiletto Visual Programmes Ltd*.

7. A clause may be unreasonable if it tries to impose unreasonable or onerous costs or procedures on the other side – *Stag Line Ltd* v. *Tyne Ship Repair Group Ltd*, *AEG (UK) Ltd* v. *Logic Resource Ltd*, section 13 of the Act.

8. Simply because a clause is included in industry-standard conditions does not mean that it is automatically reasonable – *George Mitchell (Chesterhall) Ltd* v. *Finney Lock Seeds Ltd*, *Walker* v. *Boyle*.

9. But a clause included in a set of 'model' industry-standard conditions that are accepted within the industry as fair to both sides is very likely to be accepted as reasonable – *Matthew Hall Ortech Ltd* v. *Tarmac Roadstone Ltd*.

10. This is even more likely where the clause has been specifically tailored to fit the actual contract under consideration – *British Fermentation Products Ltd* v. *Compair Reavell Ltd*.

11. A clause that sets out to limit liability is more likely to be reasonable than one that excludes liability – *Ailsa Craig Fishing Co Ltd* v. *Malvern Fishing Co Ltd*.

12. A clause that can be justified on a straightforward risk management basis is likely to be reasonable – *Photo Productions Ltd* v. *Securicor Transport Ltd*.

13. A clause that has been negotiated or agreed between organisations with an equality of bargaining power is likely to be reasonable – *Watford Electronics* v. *Sanderson CFL Ltd*.

14. But a clause that has produced a manifestly unfair result is likely to be unreasonable – *St Albans City & District Council* v. *International Computers Ltd* and *Britvic Soft Drinks Ltd* v. *Messer UK Ltd*. So, if this has happened, do proceed with care – *George Mitchell (Chesterhall) Ltd* v. *Finney Lock Seeds Ltd*.

11

Factors that May Invalidate a Contract

11.1 General

This chapter deals with three topics: 'mistake', 'duress' (physical and economic) and 'undue influence' by one party over another.

They have factors in common. They prevent voluntary agreement. They help to explain the theories that underlie the law. They are much loved by academic lawyers, and form the basis of many an examination question.

They are often used as arguments in disputes but seldom affect the practical or legal result. Also they are areas that the law controls very strictly. After all, the parties will often be mistaken about something relating to the contract; or someone may say something that is not completely correct during negotiations; or put pressure on the other side to accept a deal. If every time this happened the victim could simply walk away from the contract if he felt that it was turning out to be a bad deal, very few contracts would ever get made or completed. So the law will only invalidate a contract when there is a clear breach of very specific rules.

The subject is further complicated because these areas often overlap other parts of law, so that the court can apply different rules if it wishes to do so; see *Couturier* v. *Hastie* below, for example.

11.2 Mistake

Mistake is traditionally divided into three –

common mistake – where both of the parties are mistaken in the same way about something,

Using Commercial Contracts: a practical guide for engineers and project managers, First Edition. David Wright.
© 2016 John Wiley & Sons, Ltd. Published 2016 by John Wiley & Sons, Ltd.

mutual mistake – where the parties have interpreted the circumstances differently, so that they are genuinely at cross-purposes,

unilateral mistake – where one party is mistaken and the other party will probably be aware of that mistake and seeking to take advantage of it.

There are then two aspects of the law to consider – whether common law will
intervene to invalidate the contract, and then whether any equitable remedy will
also be available.

Common mistake

First, the common law position.

There will be common mistakes about minor details in the vast majority of
contracts. Anyone who has ever bought or sold an item second-hand or dealt
with a contract to install equipment on a site will have had to deal with the
problem.

The great majority of these mistakes are left to the contract and the parties
to sort out. The loss must then lie where it falls.

The law will only intervene to make a contract void where the mistake is so
serious that it completely destroys the deal. Therefore, actionable common mistake will only be concerned with very major issues, such as whether or not
something that is necessary for the contract does actually exist, or whether
some fundamental quality of the subject matter is present, or the ownership of
the subject matter.

> *Couturier* v. *Hastie* (1856) involved a contract for the sale of a cargo of grain
> on board a ship. Before the contract was actually made the grain had begun to
> deteriorate and the captain of the ship, as he was fully entitled to do in the circum-
> stances, had sold the cargo for what it would fetch. So the cargo no longer existed
> at the date of the contract. The buyer refused payment and the court accepted
> that there was no contract.

But the judge did not refer to the doctrine at all. He simply said that as the
cargo was specific goods and it did not exist there could not be a contract to
sell it. See now section 6 of the Sale of Goods Act.

Another example is *McRae* v. *Commonwealth Disposals Commission* (1950);
see Chapter 9 for the facts. Again there was a contract for the sale of something that did not exist. However, when McRae claimed damages and the CDC
defended itself on the basis that there could be no contract because of a common mistake, the court refused to apply the doctrine. Instead it held that as
CDC had guaranteed the existence of the ship they were liable for breach of
contract.

The best examples of the successful application of the doctrine are the
following –

Barrow Lane & Ballard Ltd v. *Philip Phillips & Co Ltd* (1929) This case concerned a contract for the sale of 700 sacks of groundnuts held in a warehouse. Neither side realised that, at the date of contract, over 100 bags had been stolen. The buyer refused payment and the court held that as 'in commercial terms' the goods had actually ceased to exist because there was no such thing as a parcel of 700 bags of groundnuts at the warehouse, the contract was void by reason of a common mistake of the parties. (The result in practice was that the seller, not the purchaser, was responsible for the cost of the stolen bags.) This principle was then followed in the case of *Associated Japanese Bank (International) Ltd* v. *Credit du Nord SA* (1988); see Chapter 9 for the facts.

Peco Arts Inc v. *Hazlitt Gallery Ltd* (1983) P bought a drawing from H. The contract stated that the drawing was an original, by Constable. Later it was discovered that the drawing was in fact not an original but a reproduction. The court declared the contract void for common mistake.

The doctrine has also been applied to cases where it was found after the contract that the seller did not actually own the goods, or that the purchaser already owned what he was trying to buy.

But where the mistake concerns not the existence or ownership of the subject matter of the contract, but some quality that it does or does not have, even if that quality is of great importance to the contract, the doctrine will not apply –

Bell v. *Lever Brothers* (1932) B was employed by L to manage part of its operations in Nigeria. L restructured its Nigerian operations and terminated B's contract, paying him a considerable sum in compensation. Later L discovered that B had been trading on his own account, which was a technical breach of his contract, although B had not realised that this was so. As a result L could have terminated his contract without compensation. L sued B for the return of the money, both for fraudulent misrepresentation, and for common mistake.

Of course this was a commercial contract, not a consumer contract, but there was probably more than a hint of a consumer (or employee) protection in the court's decision. The court threw out the fraud claim and decided that even though the subject matter of the termination settlement (B's employment contract) was valueless, the mistake was not so serious that the settlement was void.

Second, the position in equity.

Will equity step in to allow a contract affected by a common mistake about the quality (rather than the existence) of the subject matter to be rescinded, if the contract is fundamentally different to what the parties had believed it to be? In a case in 1950, a lease was affected by a common mistake.

Solle v. *Butcher* (1950) The premises were actually subject to rent control under the Rent Restrictions Act, so that the legal rent was less than £150 per annum. However, the parties had agreed a rent of £250 per annum. When the mistake was discovered the tenant claimed that the rent should be reduced to the legal amount. The court held that the common law doctrine did not apply, but used its equitable powers to rescind the lease. The landlord could now negotiate a new lease at the correct rent. (The actual problem that faced the court was a misuse of legislation. Rent control legislation was introduced after the Second World War to prevent landlords charging excessive rent for housing that was in short supply as a result of bombing. But in this case the landlord had carried out major work to the premises, making improvements and repairing war damage. He was entitled to apply for a considerable increase in rent above £150 per annum, but the tenant was trying to use the technicalities of the legislation to avoid paying a fair rent by keeping the old lease in place at a much reduced rent.)

In a later case, *Great Peace Shipping Ltd* v. *Tsavliris Salvage (International) Ltd* (2002), a ship had been chartered on an emergency basis to go to the assistance of another ship that was in difficulties, both sides believing that the two ships were 40 miles apart. They were actually about 400 miles apart, and other ships that were much nearer also stood by. An emergency charter meant that the ship had to interrupt its voyage to go to the ship in trouble and stand by her until the emergency was over. GPS claimed payment for the minimum five days charter. The court refused to set aside the charter for common mistake. It was clear that the ship could have given emergency assistance for several days if needed. So the contract was valid. The court also refused to rescind the contract under equity, on the basis that a mistake as to the quality of the subject matter was not enough to change the contract fundamentally. GPS was entitled to payment.

So equity follows the common law position.

A common mistake will only be sufficient to invalidate the contract where it relates to the existence, commercial or otherwise, or ownership of the subject matter of the contract.

Mutual mistake

A mutual mistake puts the parties at cross-purposes. One party will believe that the contract is for one thing, when the other will believe that it is for another. The result is that while they may appear to have reached clear agreement, they have actually failed to agree at all. The classic example of the problem is the old case of *Raffles* v. *Wichelhaus* (1864) in which an agreement for the sale of cotton being carried from India to the UK by a ship named *The Peerless* was invalidated by the fact that, unknown to the parties, there were two ships of that name, sailing at different times. The seller was shipping the goods on the later (slower) ship, but the buyer had expected delivery by the earlier (faster) ship. The court held that the mistake prevented the parties reaching agreement on clear terms of contract.

However, where the mistake merely relates to the quality of the subject matter then it will not prevent the parties reaching agreement. Therefore, as with common mistake the contract will be valid. The case that is usually quoted as an example of this is –

Smith v. *Hughes* (1871) H bought a consignment of oats in a crowded market from S to use as feed for his horses. Oats for feeding horses need to be matured before they can be used, because newly harvested oats will, at the very least, give a horse severe indigestion. H bought the oats after inspecting a sample and believing that they were (and that S had said that they were) 'good old oats', as opposed to 'good oats'. S, on the other hand was clear that he had told H that they were merely 'good oats'. The court was satisfied that this was a misunderstanding and held that the parties had made a mutual mistake, but they had agreed that H was to buy a specific consignment of oats at a fixed price. The age of the oats was obviously important, but only in terms of one of the uses of the goods, and so not important enough to invalidate the contract.

But this is an area where the commercial world may sometimes draw fine distinctions between different categories of goods. In a later case in contrast –

Scriven Bros & Co v. *Hindley & Co* (1913) In this case the distinction was between high-quality hemp, usable for ropes etc., and low-quality hemp, known as 'tow', which could only be used for caulking seams in ships. The total difference in usage was quite sufficient to make the subject matter of the contract different, so that the contract was held to be void for mutual mistake.

Again equity will follow the common law position.

Unilateral mistake

Unilateral mistake happens when only one of the parties is mistaken. Again if the law is to intervene the mistake must relate to something that is of serious importance for the formation of the contract.

Minor mistakes should be left to the contract to deal with. If the other party is not aware of the mistake then it would be wrong to terminate the contract. That would simply allow the mistaken party to avoid liability under the contract, at the expense of another who has done nothing wrong except to enter into a contract with someone who has not taken proper care.

So, a unilateral mistake will only be actionable when one party is mistaken about an important aspect of the contract and the other party is aware of that mistake, and is trying to take advantage of it.

Unilateral mistakes come in two different types. There may be a mistake about the terms of the contract. There may be a mistake as to the identity of one of the parties to the contract.

Mistake as to the terms of the contract

The actual number of cases in which a successful claim for termination has been made because of mistake is extremely small. Mistakes are fairly unusual anyway, and even when a mistake has been made any dispute will normally be settled before it goes to court.

A good example is the following –

Hartog v. *Colin and Shields* (1939) The case concerned a clerical error made by CS in an offer leading to a contract for the sale of rabbit/hare skins. In the trade the price of skins was normally quoted as a price per skin. By mistake the price was stated as a price 'per pound' instead. The actual weight of a skin was somewhere between a third and a half a pound. H accepted the offer and then tried to enforce the contract. (It became clear that H had realised from the start that the mistake had been made.) CS defended the claim on the basis that it was customary within the fur trade to quote prices per skin so that H must have realised that the offer was wrongly stated. The court decided that there was a unilateral mistake which concerned a material element of the contract. H must have known of the mistake and was deliberately taking advantage of it. The contract was therefore held to be invalid.

Hartog's case does however raise another point that is very important in commercial contracts.

It is clear that if one of the parties made a unilateral mistake and the other party knew of that mistake and was seeking to profit by it, then the doctrine of unilateral mistake would apply. But it will very often be difficult for the party who made the mistake to produce positive evidence that the other side did in fact know of that mistake.

What will be the position if there is no clear evidence of knowledge by the other party?

The position appears to be that a unilateral mistake will only be actionable when it can be shown that when the contract was made the other party did know of the mistake, or, because of professional expertise or knowledge, ought to have known of the mistake.

This requires a difficult value judgment. The courts have approached the question on the basis that if one party could have suspected that a mistake had been made, but deliberately said nothing, then he must be presumed to have known or suspected that the other party had made a mistake and intended to take advantage of it. See *Centrovincial Estates plc* v. *Merchant Investors Assurance Company Ltd* (1983) and *O T Africa Line Ltd* v. *Vickers plc* (1996).

Mistake as to identity

This usually happens as a result of deliberate deception. Typically, an innocent owner will be tricked into giving a thief possession of an item of value. The thief will then have sold it to a second innocent party, a buyer, and disappeared with the proceeds, leaving the two innocent parties to argue about which of them should carry the loss.

The decision in theory is simple. Did the thief have the right to sell the thing to the innocent buyer? If he did then the innocent buyer will now own it and the innocent owner will suffer the loss. If, on the other hand, the thief did not have the right to sell it, then it will still be the property of the innocent owner, who can now reclaim it from the innocent buyer, who will suffer the loss. So everything turns upon what happened between the innocent owner and the thief. If the innocent owner was tricked into giving ownership of the item to the thief, then the thief can sell it.

Put another way, if the unilateral mistake was sufficiently serious to prevent the victim entering into a contract with the thief, then the thief will not have obtained any valid title to the item and will not therefore be able to pass a good title to the innocent buyer. If the mistake was not serious enough to prevent a contract being made, then the thief will have obtained title to the item and will then have been entitled to sell it.

But there is another and opposing argument. The innocent buyer has done nothing wrong. He has not given something of value away stupidly, or acted fraudulently. Why should he have to suffer?

Of course, the contract will be voidable by the victim for misrepresentation once he has discovered the true facts, but by then it will inevitably be too late to prevent the resale of the goods.

Consider the following –

Cundy v. *Lindsay* (1878) B perpetrated a fraud on L by using a similar name and address to Blenkiron, a reputable company in the textiles trade. L sold B a quantity of linen goods, believing that it was selling them to Blenkiron. B did not pay for the goods, but immediately resold them to C. B was later convicted for fraud. L then sued C in conversion (tort) (for using L's goods as if they belonged to C) for the value of the goods. The court decided that L had been deceived into thinking that its goods were being sold not to B, but to Blenkiron, and was acting under a unilateral mistake as to B's identity. The result was that no valid contract had been made between L and B. Therefore, the goods remained the property of L, so that C was liable to pay L for them twice over.

King's Norton Metal Co Ltd v. *Edridge Merrett & Co Ltd* (1897) W used false stationery to pass himself off as a substantial company called 'H & Co' (in reality non-existent). He bought a quantity of goods from KNM and then resold them to EM. The court held that the facts were different to those in *Cundy*'s case. In *Cundy*'s case there were two separate entities, B and Blenkiron, and L had been deceived into thinking that it was dealing with one when it was in fact dealing with the other. Here there was only one entity, W/H & Co. Obviously KNM was affected by a unilateral mistake as to the attributes of W/H & Co, but there was doubt that KNM intended to enter into a contract with W/H & Co. W therefore did have a valid contract with KNM and so was entitled to resell to EM. Therefore, KNM had to stand the loss.

These two cases are typical examples of the commercial application of the rule.

The early cases took the very subjective position that everything depended on what the innocent owner believed. If he thought that the thief was a different person, such as a film actor or a wealthy man who lived nearby, then the owner did not intend to give the item to the thief. If on the other hand he simply believed that the thief was a lot richer than he actually was, then the owner did intend to give the item to him.

This was certainly the decision in the case of *Ingram* v. *Little* (1961), in which three elderly ladies selling an expensive car they owned checked the identity that the thief had used by simply looking up the name in a telephone directory. The court held that they intended to sell to the name in the directory.

Of course, the problem with this was that the innocent buyer almost invariably lost out. The only case where he did not was *Phillips* v. *Brooks Ltd* (1919). N bought/obtained a valuable ring from P, a jeweller, by pretending to be a local wealthy man. P also looked up the name in a directory. But here the court held that P intended to sell to the person in his shop, N.

The modern position has changed. The courts now take a much more objective view. If an innocent owner sells an item to another face to face, then he will be presumed to have intended to sell to the person in front of him unless there is very strong evidence to the contrary. In *Lewis* v. *Averay* (1973), a thief pretending to be a well-known television actor bought a car. The court held that the sale to the thief was valid. Although the agreement had been obtained by fraud, it was still an agreement. So the innocent buyer obtained ownership.

The current law was then confirmed in the following case –

Shogun Finance Ltd v. *Hudson* (2004) A thief stole a wallet from P and then used P's driving licence etc. to buy a car from a dealer on hire purchase. He completed an agreement form with SF in P's name. SF did a credit check on P, which showed P to be an extremely good risk, and then approved the purchase. The thief took the car and sold it to H. The court held that SF clearly intended only to deal with P, not with the thief. There was no face-to-face element in the hire-purchase agreement, so there could be no question of SF dealing with the thief. But the court confirmed the correctness of the objective approach in face-to-face contracts.

So the basic approach of any commercial organisation has to be that we do not deal with any person or organisation that we do not know on a face-to-face basis, and if we are dealing with any person or organisation that we do not know, then we do not part with goods until we have proper independent security for payment.

As a final comment on the subject of mistake, perhaps the problem for the future is sales by electronic means over the internet.

11.3 Duress

In the commercial world the use of bargaining pressure is a fact of life. It is completely legitimate and is used constantly. Every purchasing department will use buying power to get concessions from suppliers. Every seller with a dominant market position or product will use it to obtain a good price.

Law recognises this. The accepted approach to an infringement of a patent for instance is to offer to enter into a licence with the transgressor on appropriate terms, the threat being that if the offer is refused then legal action will follow. The attitude of monopoly law is that there is nothing wrong in having a dominant position in any market, or making use of that dominant position to obtain favourable terms of business, provided that market power is not abused.

But the law will not allow the use of excessive illegitimate pressure. This takes two forms, the common law doctrine of duress, that is, the use of threats, and the equitable doctrine of undue influence, that is, the use of illegitimate psychological pressure/persuasion.

Traditionally a person would act under duress when he was being threatened with death or injury to himself or to his family or friends, or to his freedom. In a more recent development, wrongful or illegitimate threats to a victim's economic interests have also been accepted as a form of duress.

Duress to the person

The threats must be such that the victim feels that he has no practical alternative except to give way and enter into a contract or accept a change in its terms. Once duress has been established the court will proceed upon the basis that the duress has caused the action taken by the victim, unless the other side can show that the duress has not been the cause, but that the victim has acted for other reasons.

An early example was *Cumming* v. *Ince* (1847), in which a lady who had been committed as a temporary patient in a private asylum agreed to sign away the ownership of her property under threat that she would be committed permanently unless she did so. The court decided that her agreement was void.

A more recent example is –

Barton v. *Armstrong* (1975) This was an Australian case. B was the managing director of a company. A, the former chairman, threatened to kill B if he did not agree that the company should buy back A's shares. B agreed, and executed a deed in his capacity as a director of the company to buy the shares. (There was in fact some evidence that B would have been prepared to buy the shares anyway, but it was clear that A had made threats to B's life and also to the lives of his family, which B did take very seriously.) The court held that as B was clearly under threat of serious injury or death, the threats did amount to duress and he was entitled to have the share purchase set aside.

In the early cases the law was that the pressure upon the victim created by the duress destroyed his consent, so that there was no contract at all, and any 'contract' was in fact void.

This view has now changed and it is accepted that duress does not automatically render the contract void (except perhaps in an extreme situation, such as a victim signing a contract while the other party pressed a loaded gun to his head). Instead it renders the contract voidable, capable of being made invalid at the discretion of the victim.

This change in principle has two important consequences. First, it recognises that however serious the duress might be, the resulting contract would still be valid if the duress did not cause the victim to enter into the contract. Second, it concentrates attention upon when pressure will actually amount to duress. This must be considered in three categories: duress to the person, duress to goods and property, and duress to the business.

Also, illegitimate pressure of a lower order than duress may still qualify as undue influence, so that an equitable remedy may be available; see below.

Obviously there are few cases of duress to the person. Most cases are dealt with under criminal law anyway. The threat must be of violent injury or imprisonment in one form or another to the victim or to members of his family. It might also be the case that the threat of injury to others might qualify as duress in appropriate circumstances. The classic dialogue in many a Hollywood gangster or cowboy film, 'hand over the money or the girl gets it', would certainly be duress.

Duress to goods/property

It used to be generally accepted that duress to a victim's property is not enough to constitute actionable duress under the common law doctrine. However, in a case in 1915, *Maskell* v. *Horner*, a demand to pay a toll was enforced by an illegitimate threat to detain the victim's goods. In giving judgment the court suggested that this was common law duress to the goods.

Subsequent cases have now established that duress to goods is a part of the law of duress. Again, there are very few examples, and duress to goods can only take place where the threat or action is clearly illegitimate. A threat, for instance, to seize goods in payment of a debt can never be duress to those goods. It is merely due process of the law. And of course this extension of the law shades into economic duress in general.

Economic duress

The principles of economic duress were only established comparatively recently. It will occur where one party has used the illegitimate threat of serious damage to the business of another to force a concession under an existing contract. The first case in which economic duress was found to exist was the following –

> *Occidental Worldwide Investment Corporation* v. *Skibs A/S Avanti* (1976) In this case the charterer of two ships threatened to go into bankruptcy, so automatically terminating the charter agreements, unless the owner agreed to renegotiate a much lower rate of hire. There had been a slump in the market so that if the charters were terminated it would be virtually impossible for the owner to find another company that would hire the ships. The resulting loss of income would have made it virtually impossible for the owner to avoid going into liquidation. This fact was well known to the charterer. The court found that the threat, virtually to destroy the owner's business, constituted duress.

In another case a few years later, in 1979, *North Ocean Shipping Co* v. *Hyundai Construction Co*, a shipbuilder threatened to terminate a contract to build a tanker unless the price was increased by 10%. The customer already had a contract to charter the ship to Shell once it was built. The possible loss of the charter, together with the consequential damage to the customer's business, forced the customer to accept the increase. Again this was found to be economic duress.

There are a number of other examples in more recent years. All are based upon a serious threat to break a contract, or possibly to act in some other way which would have a serious effect on the victim's business, in order to force the victim to agree to a change to a contract that is already in existence. Two examples –

> *B & S Contracts and Design Ltd* v. *Victor Green Publications Ltd* (1984) VG was having a stand erected by BS for an exhibition at Olympia. Some of BS's workmen building the stand went on strike. BS could have used other employees to complete the work, but instead threatened to claim *force majeure* under the contract unless VG agreed to pay a price increase of well over 10% to enable BS to settle the strike. This was found to be duress. Failure to have the stand ready for the exhibition would have had a serious effect upon VG, so that VG had no realistic alternative but to accept the demand.

> *Atlas Express Ltd* v. *Kafco (Importers and Distributors) Ltd* (1989) K had a major contract to supply a large quantity (for K) of 'Christmas Gift' baskets to Woolworths. A had entered into a contract with K to deliver the goods to Woolworths on the due date. A then realised that it had seriously miscalculated the actual cost of delivering the baskets. Therefore, when the date for delivery arrived, A instructed its drivers to refuse to load K's goods unless K agreed to a price increase of over

100%. K had no alternative but to agree. It was far too late for them to arrange alternative transport. K's general manager, therefore, after appealing to A for a change of heart, agreed to pay the extra charge, but only under strong protest. Then, when payment fell due, K refused to pay the excess. A sued K for non-payment. K defended the action on the basis that this was duress, and K's defence was accepted by the court.

In fact, the *Atlas* v. *Kafco* case is a perfect example of the way that the victim should deal with an instance of duress of any kind, making it quite clear to the other side that he regards the pressure as illegitimate and is only prepared to comply under protest.

11.4 Undue Influence

The equitable doctrine of undue influence covers a loose group of different types of behaviour, including the use of relationships of trust or personal influence to persuade or bully a vulnerable individual into entering into a disadvantageous contract, and using illegitimate commercial pressure that does not amount to full-scale duress.

The extent of the doctrine is deliberately left slightly uncertain, so that it can retain flexibility to cope with new kinds of situation should they arise. While the common law doctrine of duress and the equitable doctrine of undue influence can often seem very similar, and in particular they often deal with the same type of situation, they have very different aims. The underlying principle of the doctrine of duress is to protect the victim from the consequences of the attack made upon his freedom of contract. The aim of the doctrine of undue influence is to prevent the wrongdoer from benefiting from what he has done, by the rescinding of the contract.

The law of undue influence relates almost entirely to non-commercial contracts. It deals with situations where a personal relationship of some kind has been used to persuade someone to do something that is not in his best interests. Since it provides for an equitable remedy, it is at the option of the court, and will be lost if it has not been actively sought in due time. There is no right to claim damages for any loss that might have occurred.

Illegitimate pressure

Most of the cases relate to situations where a relationship has been misused. However, there are a few cases where illegitimate pressure has been brought to bear by one party on the other. For example –

> *Williams* v. *Bayley* (1866) A young man had forged a number of cheques. A bank made threats to the father that the bank would prosecute the son for fraud unless

the father mortgaged his land to the bank as security for the money. The bank was actually fully entitled to prosecute, and at that time prosecution for fraud could result in very severe punishment, including possibly deportation to Botany Bay. The father complied but then sought to have the mortgage rescinded. The court held that the father had proved that the amount of pressure used by the bank to obtain the mortgage was so excessive as to be illegitimate, and that the mortgage should be rescinded.

Professional advice

Where the victim trusts another person and depends heavily or totally upon him for advice, then there will always be a possibility that the advice given is not wholly disinterested. The conflict of interest between the adviser and the victim automatically raises the presumption that advice will not be disinterested. So, if that advice results in a contract or other commitment between the victim and his adviser which is not in the best interests of the victim, that will constitute undue influence –

> *Lloyds Bank Ltd* v. *Bundy* (1975) B, who was an elderly farmer lacking any real commercial understanding, had guaranteed the overdraft for his son's business. The business got into further trouble and B, acting on the advice of his bank manager, whom he trusted implicitly, increased the size of his guarantee and then as further security granted the bank a charge on his farmhouse, also to guarantee the overdraft. The business failed, and the bank sought to take possession of the house in order to sell it. The farmer defended the action on the basis that he had been subjected to undue influence by the bank manager. The court began by making the very necessary statement that the vast majority of charges, mortgages and so on are totally binding. Anyone who needs to borrow money from a bank will have to comply with the bank's lending requirements and security requirements for the debt.
>
> However, where there could be shown to be a relationship between the victim and the bank of 'trust and confidence', and in this case the bank manager did acknowledge that B relied on him implicitly to advise him, then there was automatically a conflict of interest. If there was a conflict of interest, then there was a duty on the bank to advise the defendant to get independent advice. The bank, or rather its branch manager, had not done this, and so was clearly in breach of that duty. It was, therefore, contrary to public policy that the bank should be allowed to retain the benefit of the transactions that B had entered into. The court therefore set aside the charge and refused to order B to surrender possession of his property.

Special relationships

This covers both non-family and family relationships that enable one person to be able to exert a particular influence over the other. Here the law presumes that undue influence will exist unless the defendant can show that he did not exert any undue pressure.

Of course, this is an area of law where there is always the potential for a large number of cases. This is because people in normal life often do things at the request of friends or family. In particular, many small businesses arrange substantial overdraft facilities with banks. Naturally the bank will need some security for the facility, often in the form of a charge or mortgage of the only major asset that the owner of the business has, the matrimonial home. That is often owned jointly by the business owner and his or her spouse, so that both will need to complete the transaction. Inevitably then, a number of those businesses will get into financial difficulties and the banks will need to realise their security.

Personal influence

The clearest example of personal influence was in fact the first ever case of this type, which defined the law that undue influence of this kind must be unfair or improper conduct that over-rides the power of its victim to reach his or her own decision –

> *Allcard* v. *Skinner* (1887) A entered a religious order, and remained there for some years. While there, and under the influence of N, who was her mother superior and confessor, she donated considerable property and shares to the order. After some years she left the order and then five years later tried to recover her donations. The court held that as she had had no independent advice apart from that of her confessor, she had made her donations under the influence of her confessor. This did amount to undue influence. However, as A had then delayed for five years before seeking to recover her property, she had failed to act in due time. The court therefore refused to set the donations aside, so that A failed in her claim.

Personal relationships

Personal relationships potentially create a presumption of undue influence, but only where they are sufficiently close to do so. Two examples of this type of relationship need consideration.

Parent and child –

> *Lancashire Loans* v. *Black* (1934) A mother obtained a loan from a bank, which asked for security. The mother then persuaded, or more probably bullied, her daughter into guaranteeing the loan. Later the mother defaulted on the loan and the bank tried to enforce the guarantee against the daughter. The daughter defended the claim on the basis of undue influence exerted over her by her mother at the time the loan was given. The court accepted that she had been dominated by her mother and had signed the document without properly understanding

what she was signing. She had been 'advised' by her mother's solicitor, and instructed to sign by her mother. However, she had had no independent advice, and the advice by her mother's solicitor could not be disinterested.

Husband and wife –

There is a whole series of cases involving documents, usually mortgages or charges to a bank, provided as security for overdrafts or bank loans signed by both husband and wife. When the bank has subsequently sought to enforce the security, the wife has claimed that she had been persuaded to give the security by her husband, or even by the bank, in circumstances which constituted undue influence. A typical example is –

> *Bank of Credit and Commerce International SA* v. *Aboody* (1990). A was bullied, by her husband, into agreeing to a charge on their jointly owned house as surety for a bank loan to finance her husband's business. The court held that the husband did exercise undue influence over his wife and, in the circumstances of the case, the bank was aware, or should have been aware, that he had over-persuaded his wife to enter into the surety.
>
> Therefore, the bank took the surety knowing that it had been obtained through undue influence. However, the court refused to release A from the charge, so that the bank was permitted to enforce the charge against the property. The reason was that in the particular circumstances of the case, the bank loan was not in any way disadvantageous to Mr A and A, and this was understood by A. It enabled their business to continue in being. As a result, although A had been affected by a transaction resulting from undue influence, the court refused to grant equitable relief because there was no evidence that the bank had obtained an unfair advantage from the transaction.

Another example of the same type of case which is again very instructive because of the facts, and because of the court's reasoning when giving its decision, is the following –

> *Barclays Bank plc* v. *O'Brien* (1993) Barclays called for a mortgage/charge on the O'Briens' jointly owned house as security for the overdraft (£135,000) that it provided to Mr O'B's business. Mr O'B persuaded his wife to sign the mortgage by misrepresenting to her the nature of the document, pretending that it was only for a three-year charge of £60,000. In fact, the mortgage was nothing of the kind. The standing instructions of the bank were that any manager must ensure that both parties were fully aware of the nature of the document that they were signing and that each had been given separate independent advice before the mortgage was signed. Unfortunately these instructions were not complied with, and there was evidence also that the bank did know that this was so.
>
> Some years later the business ran into serious financial trouble and the bank sought to possess the house. Mrs O'B then defended the claim. She was able to show that she had signed the mortgage as a result of undue influence and misrepresentations by her husband. The court therefore refused to enforce the mortgage,

but then laid down the rules to be followed in future. Where the transaction was at first sight disadvantageous to the second person signing the document, usually the wife, the creditor, usually a bank, must be deemed to realise, or to have constructive notice, that the principal debtor, usually a spouse or partner, might try to obtain signature by some illegitimate means.

Therefore it is the duty of the creditor to use all reasonable means to ensure that the wife enters into the transaction freely and in full knowledge of its provisions. Failure to do so will prevent the creditor from being able to enforce the transaction. The correct way to avoid being caught in this constructive notice situation is to warn the wife of the risks and advise her that she should take independent legal advice, probably without her husband being present.

To bring the position up to date –

Royal Bank of Scotland v. *Etridge* (2001) RBS's procedure was to appoint an independent solicitor to advise customers. After advice from the solicitor, E signed a charge in respect of her property to support a loan by RBS to her husband's business, but in the presence of her husband. When RBS later sought to enforce the charge E claimed that she felt that the solicitor, when giving her advice, had acted very much on behalf of her husband. (This was actually one of a number of very similar cases.) The court accepted that there was undue influence by the husband. The court then expanded on the rules laid down in *O'Brien*'s case. It was perfectly possible for one independent solicitor to advise both husband and wife as to the implications of entering into a charge/mortgage as security for a loan or overdraft facility. However, if the solicitor realised, or should have realised, that there was a conflict of interest, then the solicitor must not advise both husband and wife, but advise that the wife should have another adviser. The court then laid down detailed rules about the content of the advice that should be given.

Finally, the rules of equity relating to the consequences of setting aside a transaction for undue influence are many and complex. They are outside the scope of this book.

12

Illegal Contracts

12.1 Public influence

One of the characters in 'Toy Town', a radio series for children in the 1950s, was Mr Grouser, whose first comment on everything was that 'It ought not to be allowed'.

This chapter is about precisely that, contracts or contract terms that should not be allowed to have any legal validity. Most will be self-explanatory and need no comment. The list is deliberately kept flexible, capable of being changed, added to or extended if appropriate in the interests of public policy. However, the whole area is fraught with problems.

The basic aim of the law is to prevent people enforcing or suing under illegal contracts. If a contract is illegal, the loss must often lie where it falls, especially when work has been done or money has, or has not, been paid by one side to the other. It can produce decisions that are legally impeccable but manifestly unfair. It can catch the innocent as well as the guilty. It is also unsystematic, complex, confused, confusing and inconsistent.

Also, the terminology used does not help. To simplify matters we have avoided the use of words such as 'unenforceable', 'void', 'invalid', 'nullity', 'void *ab initio*' and so on. Instead we use only 'illegal'.

12.2 Changing values

The law is always fluid, because it has to adjust to changing values and practices within society. Contracts to sell one's children, to marry for a fee, or to procure someone else to marry are, and may always be, invalid. Surrogacy, paying someone else to carry a baby, is now accepted, though one suspects that it would not have been some years ago (although no promise to hand over that

Using Commercial Contracts: a practical guide for engineers and project managers, First Edition.
David Wright.
© 2016 John Wiley & Sons, Ltd. Published 2016 by John Wiley & Sons, Ltd.

baby for adoption will be enforceable), as is paying someone to act as a marriage broker to find a suitable partner for someone. Marriage contracts agreeing the consequences of a divorce are also now valid, whereas some years ago they would not have been.

Furthermore, the courts have always been reluctant to do anything too drastic, unless there is clear evidence of an element of wrongdoing. After all, the aim of the law is to preserve the contract, particularly the commercial contract, unless the court has no choice.

Also, there is often in these cases an element of guilty party/innocent party. Every time a contract, or contract clause, is struck down, there is a risk that the consequences might be damaging to the innocent party as well as the guilty party. And hard cases, that is, unfair decisions, always make bad law.

12.3 General

The law of contract has always had to play its part in social engineering/state policy. Starting in Victorian times this has led to a series of different types of contract being declared illegal.

Initially this was done for the purpose of *upholding social values*. First were contracts 'prejudicial to family life', which covered such things as selling children, marrying for a fee and so on. Second were contracts to 'oust the jurisdiction of the courts', or to interfere with or deny justice. Finally, there were contracts contrary to accepted morality, such things as prostitution, slavery/semi-slavery and gambling. Of course, these are all largely concerned with the non-commercial world and so are not dealt with in this book.

Then as government began to use law to regulate and control trading, business and the professions, new categories of illegal contracts emerged for the purpose of *regulating commercial behaviour*. Rent control led to illegal leases. Government licencing of various trades and professions led to the creation of numbers of 'administrative crimes', with the risk of consequent illegal contracts.

Then there were *contracts to commit or to aid crimes*. Originally crimes simply covered offences against the person or property. But the modern world has developed the concept of the administrative crime, which has automatically led to vast increases in the possible variety of illegal contracts.

Finally, there was the concept of the *regulation of activity for economic purposes*. This covers two areas – the use of 'anti-competitive practices' and unreasonable restraint of trade.

The modern types of illegal contract

There can be different kinds of illegal contract.

A contract may be legal when it is made but then become illegal because of a change in law. The most obvious example of this is contracts between companies in the UK and Germany or Italy which became illegal when war was declared in 1914 and 1939. Such contracts are normally terminated under the rules on frustration, as they become impossible to perform. Another example was –

> *Metropolitan Water Board* v. *Dick, Kerr & Co Ltd* (1918) DK took a contract in 1914 to construct a reservoir. The contract contained a *force majeure* clause and was to be completed in 1920. In 1917 DK was ordered by the Ministry of Munitions, under the Defence of the Realm Act, to stop the work so that its construction equipment could be used for other purposes. The court held that the interruption to the contract was of such a nature and likely to be for so long a period (nearly two years) that the only realistic course was to terminate the contract. No extension of the completion date under the *force majeure* clause could cope adequately with the situation.

A contract may be legal but be performed in an illegal way. In such a case the position of the parties will depend upon their knowledge or awareness of or contribution to the illegality.

A contract may be legal but contain one or more provisions that are illegal. In that case, the illegal provisions may be severable or disallowed, while the remainder of the contract continues in force.

Finally, a contract may be wholly illegal. What 'wholly illegal' means in this context is that it is impossible to carry out the contract without doing something that is contrary to law. When this is so neither party can make any use of the contract to claim against the other. Then the losses will lie where they fall, unless an innocent party can set up or justify the existence of a collateral contract to aid his claim (see below).

The areas of concern for illegal contracts

The modern law is concerned with actions/contracts that are illegal because they –

- are illegal by common law or by legislation,
- are in conflict with good government/government policy at home or abroad,
- are against the public interest, typically because they are in unreasonable restraint of trade in some way,
- are against morality and/or
- interfere with or subvert the machinery of justice within the UK.

These categories overlap each other, and to a considerable extent. So what follows is simply a summary of the main areas of importance commercially.

12.4 Contracts that are illegal

Professional rules created by statute (a recent example)

> *Mohamed* v. *Aalaga* (1998) A, a solicitor, represented asylum seekers entering the UK and seeking permanent residence. To assist in contacting possible clients, he entered into an arrangement with M to assist him by acting as his agent in return for a share of any fees that A might earn. A then failed to pay. M sued and A defended himself on the basis that the agreement was illegal, because the rules of practice of the Law Society (which have the status of delegated legislation) ban the sharing of fees earned by a solicitor. The court accepted this argument and decided that as a result the contract was illegal and could not be enforced.

This case is a perfect example of the problems that can arise. The solicitor acted in breach of his professional rules and must have realised that he was doing so from the start. The agent almost certainly did not know this, and probably could not be expected to know. He was an innocent party. However, the solicitor was then allowed to take the benefit of what the agent had done, but to escape his obligation to pay the agent for his work.

Trading licences

Here the law will consider the purpose for which the licence is required. If the purpose is to protect the public, then any contract that is in breach will be illegal. If the licence has another purpose the contract will be valid.

> *Re Mahmoud and Ispahani* (1921) A contract was made for the sale of linseed oil. At the time linseed oil traders had to be licensed under the Seed Oil and Fats Order of 1919, to protect customers against the risk of adulterated oil getting into the food or pharmaceuticals distribution chain. M, the seller, was properly licensed. As he had not dealt with I before, he asked him whether he was also a licensed trader. I said that he was licensed, which was untrue. M therefore made the contract. The price of linseed oil then fell sharply and I refused to take delivery. M sued for breach. I defended himself on the basis that he was not licensed. The court declared that the contract was illegal, so that M, who was in fact not to blame, lost his contract and his case.

But where the licence is not to protect the public, but for another purpose, the contract will be valid. The example often quoted is that of a very early case *Smith* v. *Mawhood* (1845). Dealers in tobacco were required to be licensed, on pain of a very large fine. The licences had nothing to do with trading. The purpose was to identify legitimate traders to assist in collecting import duty. S had placed a contract with M to buy a quantity of tobacco and had paid the contract price. M failed to deliver. S sued. M claimed that the contract was illegal because he was not licensed. The court decided that the licence requirement had no effect on the validity of his contracts and awarded damages to S. (A modern example might be contracts made by someone who had not registered for the purposes of paying VAT.)

In a later case the court suggested another approach –

> *Strongman Ltd* v. *Sincock* (1955) Sincock, an architect/developer, contracted with Strongman, a builder, to carry out repairs to two houses. Strongman enquired about building permits and Sincock confirmed that he would obtain all the permits needed. (During the post-war period, strict permits were required for all build- ing work so that resources could be directed to essential work.) He then did not obtain permits so that much of the work was in breach of the regulations and so illegal. The court decided that Strongman was not therefore entitled to be paid. However, the court also decided that, in telling Strongman that he would obtain permits in return for Strongman entering into the contract, Sincock had made a collateral contract that he would obtain those permits. He had failed to do this, so that Strongman was allowed to recover damages for the breach of the collat- eral contract. And the damages that Strongman was entitled to recover equalled a reasonable amount for the work that he had done.

This is a good example of the court deliberately stretching a point to produce a fair result. The developer sold the two houses at a very high profit, which was even higher because he had refused to pay the builder. The builder was com- pletely innocent; the developer was very definitely not. The collateral contract was a nice solution, and is now there for the future – but was there really any proper consideration?

Compare also *City and Westminster Properties (1934) Ltd* v. *Mudd* (1959) (see Chapter 6 above for the facts) where again the court was prepared to use the device of a collateral contract to achieve a fair result in a situation where a statement was made that procured a financial advantage.

Contracts concerned with the commission of a 'wrong', a crime or a tort

Here illegality is quite separate from the fact that the contract is usually a con- spiracy, either a crime or tort in its own right. This category includes contracts to combine in committing a crime, or to pay someone to commit a crime, or to commit a tort, such as assault, or fraud. Usually of course the policy consider- ations are quite clear.

> *Scott* v. *Brown, Doering, McNab & Co* (1892) This case concerned a contract to rig the stock market, which actually failed when the stockbroker, BDM, cheated S, the paymaster. When S sued to recover his money, his claim was rejected. The court held that as both parties knew from the start that what they were doing was defrauding members of the public into buying worthless shares, neither party could use the contract to make any claim against the other.

Contracts to defraud a public body

This category includes contracts to defraud the Inland Revenue or any other national or local authority or to evade the payment of taxes. In one case, *Napier v. National Business Agency* (1951), N was employed as a travelling representative by NBA at a very low salary, so that he was virtually exempt from paying income tax. However, his contract then provided for the payment of weekly 'expenses', at a level that was roughly six times what his actual expenses could ever be. When he sued NBA for failing to pay him, his contract was held to be illegal and his claim failed.

Contracts to benefit from the results of a crime

An example that was extremely topical at the time –

> *In the Estate of Crippen* (1911) The question arose whether Dr Crippen's estate, following his execution for the murder of his wife, included property left to him by her in her will. The court said 'no'. It stated a basic principle, 'No person can obtain or enforce any rights resulting to him from his own crime, neither can his representative claiming under him obtain or enforce any such rights'.

There are other examples of the same principle relating to cases where the murderer had insured his or her victim before committing the crime. The same principle extends to cover contracts that result in a benefit from a crime committed by another person. The clearest, and most difficult, examples of this principle arise in the world of insurance. The following is a case involving life assurance/suicide (until 1961 suicide was a crime) –

> *Beresford* v. *Royal Insurance Co Ltd* (1937) In this case, typical of the kind, the court held that the deceased's relatives, named as beneficiaries under a policy of life insurance, could not claim under the policy, as they would be benefiting from a crime, even though the deceased did not commit suicide until several years after entering into the policy. However, the court itself expressed unhappiness with the decision, 'There are many statutory offences which are the subject of the criminal law … but which … afford no moral justification to a court' to find the contract illegal.
>
> Because of course the insurance contract was not illegal when it was made.

12.5 Contracts that are legal but that are carried out in an illegal manner

A contract that is performed in an illegal way will not be enforceable by the wrongdoer if the wrongdoing goes to the heart of the contract or prevents the other party from gaining the benefit of that contract. Typical examples of this are contracts under which goods are supplied without the correct labelling, so that they cannot be resold or used; see for example the case of *Anderson Ltd v. Daniel* (1924).

However, this is rare. Normally the illegal performance will affect the performance of the contract but will not destroy the contract entirely. When this is so, the ability of a party to the contract to enforce its terms against the other will depend upon whether that party knew of the illegal performance and/or assisted the other in it.

Also, the courts are very reluctant to allow a purely 'technical' argument of illegality to be used to avoid liability.

Hughes v. *Asset Managers plc* (1995) H used AM to buy shares on his behalf. The stock market then fell sharply due to one of the Friedman-inspired recessions in the UK. H instructed AM to sell the shares again, which AM did, but obviously at a substantial loss. It was then discovered that although AM were properly licensed to deal in shares, the trader who had actually resold the shares was not. H sued AM for the loss, on the basis that they had traded in his shares in a way that was contrary to the law. The court rejected his claim on the basis that the legislation, the Prevention of Fraud (Investments) Act, was for the purpose of punishing unlicensed traders, not to unravel bad investment decisions.

Compare *Smith* v. *Mawhood* above.
For the typical approach see the following, all concerned with transport –

Archbolds (Freighterage) Ltd v. *S Spanglett Ltd* (1961) SS transported a consignment of whiskey by road from Leeds to London for AF. The contract was made on a casual basis. SS's lorry driver simply contacted AF's traffic manager to enquire whether AF had any goods that needed transport. The legislation that applied at that time allowed for two types of licence for commercial vehicles, a 'C' licence which entitled the lorry to carry goods belonging to its owner, and an 'A' licence which allowed the lorry to be used for carrying goods belonging to other people. The van used to carry the consignment only had a 'C' licence, which was known to SS, but was not known to AF. (And in fact there was no reason for AF even to question the point.) The consignment was destroyed in an accident during transport and AF claimed for the loss. SS defended the claim on the basis that because the lorry was not licensed to carry the goods the contract was illegal, so AF was not entitled to enforce the contract and the loss should lie where it had fallen. The court rejected this defence. The contract for the carriage of the consignment was completely legal. It was then carried out by SS in an illegal manner. The object of the legislation was not to interfere with the owner of any goods or his contracts for the transport of those goods, but to control those who provided the transport to ensure that they did so safely and in the correct manner. Therefore, while the contract was not to be enforceable by SS, because SS had chosen to perform the contract in a flagrantly illegal way, the contract was completely enforceable by AF. Therefore SS was held liable.

In contrast –

Ashmore Benson Pease & Co Ltd v. *A V Dawson Ltd* (1973) concerned a contract to carry two massive 25-tonne tube banks. AVD provided two articulated lorries which were not legally approved for carrying such large loads. ABP assisted in loading the lorries and their transport manager observed the work. He was aware that it was illegal to use that kind of vehicle to transport special loads of the size/weight of the tube banks but did nothing. During transport, one of the tube banks toppled from its lorry and was badly damaged. ABP sued for the cost of repairs, and the claim was disallowed by the court. By their actions, ABP had participated in the illegal performance of the contract by AVD, and so were barred from making any claim under the contract.

In another case a few years earlier the court had to deal with a less serious event of illegality –

St John Shipping Corporation v. *Joseph Rank Ltd* (1956) The Corporation had a contract to carry goods for JR. The contract was carried out properly; however, the ship was then found to have been overloaded during the voyage, so that the Corporation was fined for a breach of safety legislation. JR then refused payment of part of the price on the basis that as the contract had been carried out in an illegal way it was unenforceable by the Corporation. The court rejected this argument. The illegality did not relate to the purpose of the contract but was merely incidental to the way the contract was carried out so that it did not affect the issue. A valid contract had been made and had been carried out. The Corporation was entitled to claim payment for the work it had done.

Another variation on the theme of minor illegality –

Shaw v. *Groom* (1970) The case concerned a lease for a flat. It was a statutory requirement under the Rent Act that the tenant should be given a rent book at the effective commencement of the lease. The landlord did not do so. The tenant failed to pay the rent, and then defended himself by claiming that as he had not received a rent book when required, the lease was contrary to the law and therefore illegal. The court held that although the rent book had not been provided the lease was not illegal, so that the landlord was entitled to recover his rent. It is very clear from the decision given by the court that it was not prepared to strike down the lease for a failure to comply with a requirement that was of administrative importance but did not go to the heart of the contract.

'Trading' contracts that contain an illegal provision

Commercially, this area covers contracts in restraint of trade/competition.

The basic rule is that if a contract contains a restrictive undertaking that prevents one party or the other from engaging in his normal work or trading activities, that undertaking is likely to be unenforceable unless it can be shown to be fair and reasonable in the circumstances.

This is a difficult area of law, because a number of different areas of public policy collide with one another. It covers a number of separate but related topics –

- contracts between employer and employee;
- contracts for sole rights;
- sale of business contracts;
- tied distributor-style contracts;
- collaborative trading contracts.

(In addition to the above there is also competition and monopoly/dominant position law. Within the European Union as a whole this is regulated by statute under Articles 81 and 82 of the Treaty of Rome. Within the UK the relevant legislation is the Competition Act 1998 and the Enterprise Act 2002. The Acts consolidate previous law and introduce a number of concepts into the law from the European Union. In addition, for any organisation involved in contracts concerned with international trade, there are a number of other provisions of European Union law relating to cross-border trading contracts, intellectual property licensing, the use of commercial agents and representatives, and so on which need to be taken into account. These areas are outside the scope of this book, but the law in this area is complex, and complicated, and no organisation should proceed too far without taking proper advice.)

12.6 Contracts between employer and employee

This is an area that is commercially used every day. At any one time there will be thousands of employees and ex-employees with employment contracts that contain restraint clauses – that never cause problems.

The normal employment relationship/contract gives the employer ownership of everything that the employee makes in the course of his normal work, including whatever he writes or draws (copyrights). It also requires the employee to maintain a degree of confidentiality about his work, which is often reinforced by specific clauses in the contract. But any intelligent employee in a position of responsibility cannot help but learn information which would be of value to a competitor. What is the position if his employment is terminated?

There are two conflicting interests.

First, there is the interest of the employee. He must have the right to earn his own living using all the skills and knowledge that he possesses. That is simple economics. And of course the greater and the more specialist those skills and knowledge, the more potential value he has to his employer's competitors, and the more restricted his choice of possible work and employers. This interest of course then combines with the basic policy of protecting the individual against the bargaining power of the organisation.

The employer, on the other hand, has the right to protect his own proprietary knowledge/information, and his trading position. Obviously some knowledge may be protected by intellectual property rights such as patent or copyright, but the vast majority of commercial information and technical know-how can only be properly protected by keeping it confidential, and preventing it being given or made available to competitors.

A typical example of the situation is the following classic, and very important, case –

Morris (Herbert) Ltd v. *Saxelby* (1916) M manufactured mobile cranes. S was the manager of M's design department. S was in fact a brilliant engineer/designer and had been responsible for several significant improvements in the designs of M's products. S was also shortsighted and needed to wear glasses at all times. This was important because it meant that S was not eligible for service in the armed forces during the First World War. M had included in S's contract of employment a clause that provided that S must not work on his own account or for any competitor of M within the UK in any capacity for a period of seven years after termination of his employment with M. However, as soon as the First World War started there was a considerable increase in demand for mobile cranes, and for experts in their design. S was headhunted by a competitor, with a very much improved job offer, and handed in his notice. M immediately went to court for an injunction to restrain S from taking up his new post. The case went as far as the House of Lords, because it was recognised that there were very serious issues here, and this was almost the first case of its type in the modern era. The court held that the clause was far wider than was reasonable to protect M's interest, so that it failed and could not be enforced by M.

The court laid down the principles that still apply to clauses of this kind today.

- No restraint clause can be valid unless the employer can show that he has an interest that justifies protection.
- The interest must arise because the employee has access to special knowledge or information which the employer reasonably thinks of value to a competitor.
- The information may be any kind of commercial or technical knowledge, skill or know-how.
- The interest cannot extend to anything else except that information.
- Once that interest is proved the employer can include a restraint clause in its contract with that employee.
- The clause may impose a reasonable limit on the freedom of the employee after termination.

- This limit relates to use of the proprietary information to compete with the employer or to making it available to others to enable them to compete.
- But the clause must not prevent the employee from using his own skills and knowledge to earn his own living, nor must it demand more than the minimum needed to protect the proprietary information.
- Any clause that asks for more than the minimum will be wholly illegal.
- The other terms of the contract will be unaffected.

In a side issue that arose in *Morris*'s case the court also had to consider the right to use knowledge actually created by the employee himself. On leaving Morris Ltd, Saxelby had taken with him his personal copies of various documents that he had written setting out the methods of calculating various aspects of crane design. The court ruled that the documents contained know-how which belonged to Morris, and so Saxelby was not entitled to them.

Every restraint clause must be considered on its own merits in the light of the actual circumstances of the case. A reasonable clause will succeed in its object. A clause that fails to be reasonable in any way will be wholly invalid. Also, the clause will be considered as a whole. Attempts to avoid the invalidation of a restraining clause by splitting it into several parts, clauses or sub-clauses will usually fail. (However, where a contract contains two quite separate restraints, the court may be prepared to treat them as such. In the case of *Home Counties Dairies Ltd* v. *Skilton* (1970) a contract contained separate clauses, the first preventing the employee, a milkman, from taking any employment in the dairy industry, and the second preventing him from working as a roundsman or approaching his existing customers. The first was disallowed, while the second was accepted.)

In deciding whether any clause is reasonable, the law will usually take into account three main factors. Using the *Morris* case as an example, the three factors to be considered are as follows.

- *Scope or type of work* – S was employed as manager of the design department; however, the restraining clause would have prevented him working for a competitor 'in any capacity'. The court held that it would have been reasonable to prevent him from working for a competitor in a design capacity, but to extend this to work of any type was far wider than reasonable.
- *Area and type of business* – The area of the restraint was that S should not work for any competitor of the employer throughout the whole of the United Kingdom. The court held that in the circumstances where the other crane manufacturers were based all over the country, this was reasonable.
- *Period* – The court held that seven years' restraint was far too long to be reasonable. In effect it would prevent the employee from earning a reasonable living for himself and his family by doing the one thing that he was

qualified to do for such a long period that he could be reduced to poverty. (See the comment below.)

When considering whether a particular restraint is reasonable the court will take into account –

- the information available to the employee – the greater the knowledge the more reasonable any restraint;
- the type of knowledge – as a general rule commercial knowledge, such as customers, prices etc., will quickly become out of date, whereas technical knowledge deserves much longer protection;
- the seniority of the employee – the more senior the more severe the justifiable restraint, and so on.

There are many examples of how these considerations are applied to actual circumstances. In the case of *Fitch* v. *Dewes* (1921) a restraint area of within seven miles from Tamworth was held to be reasonable for a solicitor's clerk employed by a practice in the town. Tamworth was a small country town, so that the chance of damaging the employer's business was quite high. In a later case, *Fellowes* v. *Fisher* (1976) a restraint area of seven miles (obviously modelled on *Fitch* v. *Dewes*) for another solicitor's clerk, but in a busy London suburb, was held to be unreasonable, because the population was so much larger.

The essential rule in drafting any restraint clause seems to be to consider every case on an individual basis, only impose the minimum restraint necessary, and always leave the employee reasonable opportunity to earn his living.

This can be difficult. If more recent cases are anything to go by the more highly specialised the work done by the employee, the harder it is to devise a practical restraint that will be accepted, simply because any restraint will prevent the expert employee using the only skills that he has to earn his living. Indeed, the logical position is that where an employee is so highly specialised that he is only qualified for one type of work, no restraint clause can be valid. (An example might be a trader in stocks and shares, or a currency trader in a merchant bank.) In that case the only alternative for the employer is to include lengthy notice of termination periods in contracts, allied to a policy of 'gardening leave' where appropriate. (And even that may be contested where an employee's earning ability depends upon his being allowed to present his work to the public, such as an author or critic for example.)

As a final point on this, in a case in 1959, *Kores Manufacturing Co Ltd* v. *Kolok Manufacturing Co Ltd*, two companies who manufactured specialist products agreed that neither would employ anyone who had previously worked for the other. The court held that this provision was in unreasonable restraint of trade, because they were seeking to restrict their employees' freedom in a way which would not be permitted in an employment contract.

12.7 Contracts for sole rights, etc.

In the situation where a person is not an employee, but is still in an unequal relationship with an organisation, the same principles apply. An example might be an arrangement between an individual inventor and a company wishing to license and exploit that invention. The case usually quoted comes from the popular music world –

> *Schroeder (A) Music Publishing Co Ltd* v. *Macaulay* (1974) SMP gave a contract to M, a songwriter, on its standard terms. The contract was for an initial period of five years and gave SMP copyright in everything that M wrote during that period. If he was successful then the agreement was automatically extended for a further five years. M had no right to terminate the contract, but SMP could terminate at any time and also assign the contract to anyone else. The contract was typical of those used in the music industry at the time for young/new composers. M, who was successful, asked for the agreement to be terminated. The court held that the contract simply represented an inequality of bargaining power. Inequality of bargaining power did not mean that the contract was automatically illegal, but it must not be used to drive an unfair bargain. M's claim was upheld.

The court also commented on the defence advanced by SMP that their contract was typical within the industry. Just because it was typical did not mean that it was fair.

12.8 Contracts between equals

The private individual, when not an employee, and the organisation both have the right to use any information or knowledge that they may have to engage in any business that they may wish, unless they have agreed under the terms of a contract that they will limit their activities in some way, unless the nature of the restraint is so one-sided as to be unreasonable.

Where a person agrees to a restraint clause as between equal parties, the law will readily accept much wider restraints, provided that they can be shown to be reasonable. In the case of the sale of a business, for example –

> *Nordenfeldt* v. *Maxim Nordenfeldt Co* (1894) N was an entrepreneur and inventor within the international armaments industry. His company was highly successful, one of its chief products being a light artillery quick-firing gun, the Maxim gun, used both on the battlefield and later as an anti-aircraft weapon even as late as the Second World War. N was personally responsible for the invention and development of the Maxim gun. He decided to sell his business, and agreed to a clause in the sale contract, requested by the new owner, restraining him from engaging in the armaments business anywhere in the world for 25 years (in effect for ever).

Later N returned to the armaments business with another company. MNC sought to restrain him and the court accepted that the clause was reasonable. If N were to be allowed to compete with his former company its business might well suffer.

The protection that was to be given to the individual as an employee, to allow him to earn his living doing what he could do best, was one thing. However, here that individual was acting not as an employee but as an equal player, and it would be hard to think of a more equal player than a multi-millionaire businessman selling his own company for a very substantial sum. Equally it was not unreasonable for the purchasers of that business to want protection against competition from him. In the circumstances then, if N was prepared to accept a restraint, however wide that restraint might be, he was completely free to do so.

In the case of a joint trading arrangement –

English Hop Growers v. *Dering* (1928) This case concerned one of a whole series of agreements between the individual hop growers and the EHG, their trade association, under which each grower agreed to sell his entire crop to the EHG, which then acted as a joint marketing organisation for the entire industry. This ensured that profit/loss on each year's crop was shared equally over the whole industry, and protected small growers from exploitation by the large buyers in the brewing industry. The arrangement was held to be reasonable in the circumstances, as it was fair both to EHG and to D, and of course to the other growers as well.

Where however, the arrangement between the two parties is rather less equal and results in a restraint clause or clauses that are not reasonable, then the court may well take a different view. Typical examples of this situation appear in a number of the 'exclusive dealership' industries such as publishing, especially music publishing, as we have already seen, motor vehicle and petrol/oil distribution and so on. In one well-known case –

Esso Petroleum Ltd v. *Harper's Garage (Stourport) Ltd* (1968) HG entered into 'sole supply' agreements relating to two garages The agreements provided that the garages must remain open for the sale of petrol etc. for the duration of the agreement, laid down the opening hours, and required HG to use EP's promotional material and sell only EP's products. One was for 4.5 years, and the other for 21 years, and was coupled with a loan from EP to improve the premises backed by a mortgage as security for repayment. After some years HG started to sell petrol supplied by others, and EP went to court to enforce the agreements. The case went to the House of Lords. The Lords decided that the mortgage did not count as a contract, so that the covenants in it did not count. However, the court then considered whether the two sole supply agreements were reasonable. The court held that the provisions relating to the way HG was to run its business were onerous, but not necessarily unreasonable in the context of marketing petrol. Then the court turned to the question of whether the duration of the agreements was reasonable. The 4.5-year agreement was upheld, but the 21-year agreement was not.

It extended far longer than the period that was reasonably predictable in business terms, which had to be the deciding factor.

In the real world there are many such agreements. The guidelines are simple. All they need is a little careful thought.

12.9 Footnote: the unauthorised taking of proprietary information

Although it is not strictly related to the English law of contract, in a Scottish case, *Weir Pumps Ltd* v. *CML Pumps Ltd* (1984), the court had to deal with the unauthorised acts of employees leaving WP in taking proprietary technical information, in the form of manufacturing drawings, which belonged to WP, and giving that information to CML, their new employer. CML then used that information to compete with WP. The court held that the original taking amounted to theft, and that CML used the information knowing that it was the (stolen) property of WP and that WP had not consented to its taking. The court therefore ordered that the information, plus all copies, should be returned or destroyed, and awarded substantial damages against CML.

13

Privity of Contract

13.1 Introduction to the privity rule

The law sees a contract as a private matter between the parties (hence the term 'privity', or 'privacy', of contract). The basic rule is simple. The parties agree the contract. The contract sets out the rules and the rules can be enforced. But the rules can only be enforced by the parties to the contract. They cannot be enforced by or against anyone else.

But from time to time we make contracts for other people's benefit, not just for ourselves. In private life we buy presents or package holidays for our family. In commerce we cascade obligations and liability up and down chains of contracts and subcontracts. In groups of companies one company may buy, and other companies may use or operate. Whenever we do this, privity of contract makes difficulties.

There are three main problems –

- who can demand that the contract is carried out?
- who can claim compensation if it is not carried out – and what can they claim compensation for?
- who can claim to be protected by the contract against claims?

The answers given by the basic rule are that only the parties to the contract can require it to be carried out, or claim compensation if it is not carried out. The compensation that they can claim is substantial damages for any losses that they have suffered as a result of a breach – but whether they can claim in respect of losses suffered by others is very uncertain. Finally, only the parties can take the benefit of any liability exemption clause in the contract.

Using Commercial Contracts: a practical guide for engineers and project managers, First Edition. David Wright.
© 2016 John Wiley & Sons, Ltd. Published 2016 by John Wiley & Sons, Ltd.

Two examples of the strict application of the rule, one private and the other commercial –

Tweddle v. *Atkinson* (1861) Son T became engaged to the daughter of G. Both sets of parents obviously approved of the match, because G entered into an agreement with father T by which both of them would pay an annual allowance to T. Father T then died. G later failed to make the payments, and then also died. T sued G's executor, A, for the money (his own father clearly could no longer do so). T's action failed. The judge simply stated that there was no legal precedent that supported the principle that a 'stranger' to a contract could take advantage of that contract, even though the contract was clearly made for his benefit.

Cascading obligations along a chain of contracts –

Dunlop Pneumatic Tyre Co Ltd v. *Selfridge & Co Ltd* (1915) (In 1915 resale price maintenance was allowed.) DPT sold tyres to a distributor. The contract required that if the distributor sold tyres for less than DPT's resale list price, he should pay DPT £5 per tyre. Also, if he sold any tyres to another company 'in the motor trade' he should impose the same terms. He sold tyres to Selfridge and duly imposed DPT's terms. Selfridge then sold two tyres to one of its own customers below the list price, and refused to pay £10. This was clearly a breach of the contract between himself and the distributor, but DPT sued Selfridge for £10 directly without involving the distributor (as a test case). The court dismissed the claim. It said 'only a person who is a party to a contract can sue on it'. DPT was not a party to the contract between the distributor and Selfridge and so could not sue under it.

Finally, an example of an exemption clause in a chain of contracts –

Scruttons Ltd v. *Midland Silicones Ltd* (1962) Goods belonging to MS were being shipped. The contract between MS and the shipper, United States Lines Inc, limited the liability of the shipper for any damage to the goods to £200 per package. As was normal, USL contracted with S, a stevedore, to unload the goods. They were damaged by the negligence of S. MS sued S for damages, and S claimed that it was protected by the liability clause in the contract between MS and USL. (Again this was a test case financed by the insurance companies on either side.) S based its defence on three separate arguments. First, as it was known to MS that any shipper would automatically employ a stevedore as a subcontractor to unload the goods, it must be implied by the custom of the trade that S was protected by any limit of liability that applied to USL. Second, USL was acting as S's agent when it made the contract with MS. Finally, even if there was no term to be implied by the custom of the trade into the shipping contract, then on the basis of the officious bystander test it was to be assumed that there was an implied term protecting them as an automatic subcontractor. All these arguments were rejected by the court. The second argument, on the basis of agency, was rejected because there was nothing in the words of the contract to support it, and the shipping contract was made well before USL's contract with S. The arguments based on implied terms were dismissed because S could not produce satisfactory evidence to support them. S was therefore liable to MS for the full value of the damaged goods.

13.2 Exceptions to the rule

General

The rule of privity does have obvious advantages. It clarifies and simplifies the contractual relationship. It prevents conflicting and overlapping claims that would make insurance very much more costly and contract management difficult. But it can create injustice and problems.

The law has been trying to solve these problems ever since day one. It has done so by creating exceptions to the rule of privity, though not very successfully. It has used the law of agency, created the principle of positive and restrictive covenants applicable to land, and used the idea of subrogation. Then parliament has stepped in by creating further possible exceptions to the doctrine under the Contract (Rights of Third Parties) Act 1999.

But these exceptions are still just that. In the great majority of cases the doctrine of privity will still apply. The consequence is that it is always necessary to make the contract with the right company or companies. It is important to sort out the potential relationships before contract, when there is still time, rather than afterwards when it may be too late. Sometimes it may even be worthwhile placing the contract in the names of two or more user companies, rather than one purchasing company.

The practical solutions for the buyer are to –

- use the device of 'assignment' of rights, or
- enter into the contract on a 'joint' or 'joint and several' basis, or
- use the law of agency, or
- use collateral contracts.

Most commercial issues focus on cascading contractual obligations up and down the chain of contracts that go to make up the complex project. We may need to move 'guarantee/repair' obligations upwards from equipment suppliers. Confidentiality and intellectual property obligations may need to be moved in both directions, depending on where the technology sits. Rationalising insurance cover will require the use of liability exemption provisions. And so on.

As is always the case with common law many of the classic cases come from the nineteenth century, though the principle has been regularly applied. There are two main types of case. Someone who is not a party to a contract will use it to justify a claim independently of the party that made the contract. Alternatively, someone who is not a party to a contract will use it to defend himself against a claim on the basis that he is protected by a clause in the contract limiting or excluding liability.

Compensation claims

When a breach causes loss to a third party for whose benefit the contract has been made, instead of or as well as the actual party to the contract, can he claim substantial, as opposed to nominal, compensation for those losses? This is what happened in the following case –

> *Jackson* v. *Horizon Holidays* (1975) J bought a top-range four-week family holiday in Sri Lanka for himself, his wife and three very small children at a hotel described in the HH brochure as providing the highest standard of accommodation, at a price of £1,200 (the equivalent price today would be anything up to £20,000). At the last minute HH changed the holiday to a different hotel, because the original hotel was not complete. HH promised J that the new hotel was of the same standard. The replacement hotel turned out to be absolutely appalling, particularly for a family with a new baby and two toddlers. The rooms were insanitary, with green mould on the walls, no proper washing or toilet facilities, or even toilet paper, dirty bed-linen and so on. J complained, and the family were transferred to the original hotel but without much improvement. J sued for damages on the basis that the whole holiday been a nightmare for both himself and his family. HH accepted liability – the only dispute was about the extent of damages. It was quite clear that J was entitled to claim compensation for his own personal distress and the loss of his holiday. The question was could he also claim substantial damages for the distress etc. suffered by the other members of his family as well. The court awarded J damages of £600 for his own distress and a further £500 for the distress suffered by his wife and family, 'as it was only fair' to do so.

Clearly in Jackson's case the court accepted that J, who made the contract, was also entitled to claim full compensation for losses suffered by his family. This seemed to solve the problem and allow substantial damages to be awarded for losses caused to a third party, where the contract was made for their benefit.

However, the decision in Jackson's case was questioned in later cases, in particular in *Woodar Investment Development Ltd* v. *Wimpey Construction UK Ltd* (1980). The case arose out of a contract between Woodar and Wimpey under which Wimpey was to pay a sum of money to a third party, actually a company owned by the original seller of land to Woodar. Wimpey failed to do so, but the court held that Woodar were not entitled to sue for damages for the non-payment unless they had also suffered loss, or were acting as agent or trustee for the third party (which Woodar could not do).

Another case at about the same time was *Forster* v. *Silvermere Golf and Equestrian Centre* (1981). F transferred a valuable parcel of land to the Centre in return for an undertaking that the Centre would build a house on the land for her and her children to occupy rent-free for their lifetimes. The Centre failed to do so. F sued but was allowed compensation only for her own loss. The court refused compensation for the loss suffered by her children.

See now section 1(5) of the Contract (Rights of Third Parties) Act 1999, and below.

The exceptions are all created to remedy perceived injustices that arise from the rule.

Use of the rules of agency – agency created by contract

If the person who enters into the contract does so as agent for the person who seeks to enforce its terms then the doctrine of privity simply does not apply.

In the *Midland Silicones* case the court was not prepared to find a way round the rule by implying terms into the contract that would allow USL to act as an agent for his subcontractors. But in the following case, a few years later –

New Zealand Shipping Co Ltd v. *A M Satterthwaite and Co Ltd* (1975) This case concerned goods being carried by a ship called *The Eurymedon* from the UK to New Zealand. The facts were similar to those in the *Midland Silicones* case. But the contract contained the following very comprehensive wording:

It is hereby agreed that no servant or agent of the shipper (including every independent contractor from time to time employed by the shipper) shall in any circumstances whatsoever be under any liability whatsoever to the ... owner (of the goods) ... and ... every exemption limitation condition and liberty herein contained ... shall ... be available and shall extend to protect every such servant or agent of the shipper ... and for the purpose of all the foregoing provisions of this clause the shipper is or shall be deemed to be acting as agent or trustee on behalf of and for the benefit of all persons who are or might be his agents from time to time (including independent contractors as aforesaid).

One of the clauses of the contract then provided that no claims for damage to the goods would be valid unless brought within 12 months after delivery to New Zealand. Goods were damaged by a stevedore while being unloaded in New Zealand, but the action was not brought within the 12-month period. Was the stevedore protected by the 12-month limitation period? The court decided that he was. This was a commercial contract, obviously negotiated between parties who knew precisely what the risks were and who were mainly concerned with who should be responsible for insuring the goods. The court accepted that if the parties were agreed that the contract should include such a clause then it was not for the court to argue. The clause provided that the shipping company was acting as agent for his subcontractors, and the stevedore's defence to the claim succeeded.

The clause used by the shipping company in the *Eurymedon* case was very carefully drafted and very comprehensive. It specifically included actual and future subcontractors, as well as employees and agents. It created an equitable trust as well as an agency. It referred to every type of contractual benefit and protection. And so on. It is clear from the judgment of the Privy Council that the court accepted that it was virtually watertight. It is uncertain what the position would be if a contract included a less comprehensive clause. But it is probable that the clause would still succeed. It would simply need to be carefully drafted and accepted by the other side.

Use of the rules of agency – agency created by statute

Under section 56 of the Consumer Credit Act 1974 a statutory agency has been created in respect of 'regulated contracts', that is credit sales, and particularly hire-purchase contracts, entered into by consumers. The hire-purchase transaction generally consists of two transactions, a sale by the dealer to the finance company, followed by a hiring contract between the finance company and the buyer. Before the Act, the position was that the dealer acted only as an agent for the buyer to set up the hiring contract. So, as there was no contract between the buyer and the dealer, the only binding statements made by the dealer to the buyer were those relating to the description of the goods being sold.

Under the Act the dealer will now also act as an agent for the finance company. This means that the finance company is now responsible for everything said by the dealer to the buyer before contract.

See also the case of *Brown* v. *Sheen & Richmond Car Sales Ltd* (1950) below. The consequence of section 56 and *Brown*'s case is to give a buyer the right to sue the dealer for damages for breach of a collateral warranty (under *Brown*'s case) or to sue the finance company under section 56 for damages/reduction of payments or even for cancellation of the hire-purchase contract if the facts justify it.

Another possibility is that the party may be able to demonstrate that he falls within the definition of an undisclosed principal (see Chapter 2), although this is rare.

Land law

There is a further exception that applies in the case of contracts that create a 'positive covenant', a permanent interest attaching to land. This was decided in the following case –

Smith and Snipes Hall Farm v. *River Douglas Catchment Board* (1949) In 1938 the Board entered into an agreement under seal with a group of landowners, who between them owned a large area of land bordered on one side by the raised bank of the Leeds–Liverpool canal and on the other side by a river. The river was prone to flooding and the land was virtually useless except as seasonal grazing land. The

Board proposed, in return for substantial payments by the owners, to raise the riverbank and then to maintain it 'for all time'. For a few years the scheme worked, and the land was transformed into valuable agricultural land. But in 1946 there was a serious flood caused by a failure of the bank and the year's crops were lost. In 1940 one of the original landowners had transferred her land to S who had then leased it to SHF. They now sued to recover their losses. The court held that where a party to a contract, in this case the Board, entered into an undertaking to carry out work in relation to land which affected the value of that land, the undertaking would become a positive covenant which was attached to the land. As a result it would be able to be enforced by whoever owned the land or held an interest in it. Therefore both S and SHF were entitled to recover.

Transfer of the right to sue (in another's name)

Subrogation, 'substitution', is not really an exception to the doctrine of privity at all. The term relates to a device used within the insurance world. Where an insurer has agreed to indemnify an insured against loss or damage to any asset or goods, the policy may reserve a right of subrogation to the insurer. Then in the event of any claim by the insured which is the result of damage caused by a third party, the insurer may meet the claim and then claim reimbursement from the third party, in the name of the insured. Of course, most claims will be made under the law of tort for negligence by the third party, but may also involve claims for breach of contract. For an example see *Photo Productions Ltd* v. *Securicor Transport Ltd* in Chapter 10 above.

Dual capacity

This concerns claims made other than in a purely personal capacity, in effect where the claimant can show dual capacity. This exception is illustrated by a case similar to *Tweddle* v. *Atkinson* –

Beswick v. *Beswick* (1967) Uncle B, who was elderly and in poor health, entered into an agreement with nephew B to transfer his business to him, in return for a consultancy agreement for his lifetime at a fee of £6.10 shillings a week, together with an annuity for his widow, aunt B, at the rate of £5 a week for her life. After uncle B's death the nephew paid one £5 payment and then stopped. Aunt B sued for the arrears and also for a court order for specific performance ordering nephew B to continue paying the annuity. As she had been granted letters of administration for her husband's estate, she sued both in her personal capacity and as his admin-istratrix, and was successful. The court decided that she had no right to claim in her personal capacity. But she was successful in her capacity as an administratrix representing her late husband. The court also decided that in this case damages would not be an adequate remedy. As uncle B could suffer no loss from nephew B's breach of the agreement, only nominal damages could have been awarded.

> It would be unfair to deny an order of specific performance, as this was the only remedy that would ensure that the agreement was properly carried out.

In both *Smith and Snipes Hall Farm* and *Beswick*, the courts were determined to find a way round the doctrine of privity. Both cases had arisen from shoddy practice by the defendant. In *Smith and Snipes Hall Farm* the Board had made a mess of its costings and therefore done a cheap job that in the opinion of one expert witness (and of the Board's own engineer) was doomed to failure from the start. In *Beswick*'s case the nephew's behaviour was even worse.

Insurance arrangements

There are a number of minor exceptions to the doctrine that for very practical reasons have been created by the insurance industry. These relate to life assurance where the beneficiaries of a policy need to be protected, motor insurance when a vehicle is driven by someone other than the owner but with his consent, and property insurance against fire damage. The problems of *Jackson*'s case, above, are now usually covered by insurance as well.

Statutory rules

Compulsory professional indemnity insurance – under section 37 of the Solicitors Act 1974 the Law Society has established and operates an insurance scheme which creates reciprocal rights between insurers and solicitors. The scheme operates as a part of statutory public law, for the protection of members of the public against loss that they might otherwise suffer from dealings with the profession, not as a private contractual arrangement run by a trade association for the benefit of its members. The result is that the right of any person to claim under the scheme is independent of any contract with a member of the profession.

Occupiers of premises – under the Occupiers Liability Act 1957 any occupier of premises who has a contract that requires him to allow third parties to enter those premises will be liable for any injury to those third parties if the premises are unsafe, whether or not they were parties to the contract under which the right to enter the premises was been created. This is again an example of a statutory duty imposed to bypass any possible effect of the doctrine.

13.3 Other solutions

The difficulties caused by the doctrine have led to a number of attempts to find ways round them. Most, such as the use of trusts, are outside the scope of this book, but there are two that deserve mention.

The 'collateral contract'

A collateral contract is one that runs alongside another contract, or in this case a chain of contracts. One example is what is often called a 'subcontractors warranty, often used in building/construction projects. This is a formal collateral contract between a significant subcontractor of whatever tier and the end-client guaranteeing that any work or equipment supplied will be of good quality and in accordance with the subcontract. This agreement serves two purposes. First, it ensures that the client has a direct link with the suppliers of the various items of equipment so that when those items need maintenance or repair he can deal directly with the supplier. Second, if equipment turns out to be totally inadequate then the client has a direct claim against the supplier.

> *Shanklin Pier Ltd* v. *Detel Products Ltd* (1951) SPL placed a contract with a contractor to paint the pier with high-quality anti-corrosion paint. DP then approached the managing director of SPL with a proposal that the pier should be painted with DP's 'DMU' paint, which DP warranted to have a life of seven to ten years. DP convinced both the managing director and architect of SPL to use DMU paint, and SPL therefore instructed the contractor to use DMU paint. The contractor did so and applied it to the pier. His work was carried out to a proper professional standard. Unfortunately the life of DMU paint turned out to be not seven to ten years, but rather less than six months. SPL could not claim against the contractor. He had carried out his work properly and used the paint specified by SPL and therefore could have no liability. SPL therefore sued DP on the basis that by promising that if DMU paint was used it would have protected the pier for at least seven years, DP had made a contractual offer which was accepted by SP when it instructed the painting contractor to use the paint. The claim was successful. The court said 'I see no reason why there may not be an enforceable warranty between A and B supported by the consideration that B should cause C to enter into a contract with A, or that B should do some other act for the benefit of A'.

The judgment in the *Shanklin Pier* case is short. The judge was clearly satisfied from the evidence that Detel Products had made categorical promises to the pier management company regarding the protection that would be given to the pier by its product, and that these promises had caused the pier company to vary its contract with the painting contractor. For this reason the judge was prepared to accept that there could be a separate contract made between the parties running alongside the main contract–subcontract chain.

The collateral contract principle has been extended in a number of other cases following the *Shanklin Pier* case. In the *Shanklin Pier* case, the promise made by DP was in relation to a specific transaction/project. In a case in 1965, *Wells (Merstham) Ltd* v. *Buckland Sand and Silica Company Ltd*, in generally similar circumstances a promise was held to be capable of forming the basis of a collateral contract even though at the time it was made there was no specific purchase contract in view. This is in fact extremely important because it means that 'over-selling' may create collateral contracts. The principle has also been extended to include promises made about conduct by one of the parties –

> *Charnock* v. *Liverpool Corporation and Another* (1968) C's car was damaged in a collision with an LC bus, due to the negligence of the bus driver. C took his car to a garage for repair, and the cost of the repair was agreed between the garage and LC's insurers. In normal circumstances the maximum time required to repair the car would have been five weeks. In actual fact the repair took over eight weeks. C had hired a car while his own was being repaired. LC's insurers refused to pay C's car hire costs for more than five weeks. C sued the garage for the extra car hire and won his case. The garage had known all along that the repair might be delayed due to the pressure of other work, but had made no mention of this to C. The court decided that by not telling C that repairs to his car might be delayed, the garage had in fact promised C that in return for his giving them the opportunity to repair his car and to be paid for doing so by LC's insurers, the repair would be carried out within a reasonable time.

The principle of the collateral contract has also been applied to hire-purchase contracts; see above –

> *Brown* v. *Sheen and Richmond Car Sales Ltd* (1950) B purchased a second-hand car from SRC under hire purchase. Before the purchase SRC told B that the car was 'in perfect mechanical condition'. The car then turned out to be in very poor mechanical condition and B had to spend a considerable amount of money in making the necessary repairs. He claimed damages from SRC on the basis that this statement was a collateral warranty, and his claim was successful.

Claims in negligence

The typical situation will then be that a subcontractor will have acted in a way that has caused loss to the client, and for some reason the client does not wish or is not able to claim from the contractor.

There are two possible situations. The subcontractor may have caused damage to the property of the client, and can be shown to owe a duty to the client to take reasonable care. A good example of this is the following –

> *National Trust* v. *Haden Young Ltd* (1995) (The Uppark fire) The Trust was carrying out extensive repairs to the roof of Uppark, a major stately home. The Trust employed a local company as main contractor and the main contractor employed HYL to work on the lead cladding on the roof. During a tea break, employees of HYL left a blowtorch burning in the roof. This caused a fire which resulted in major damage costing in excess of £20 million to repair. The main contractor did not have anything like enough insurance cover to meet the cost, and anyway went into liquidation soon afterwards. Therefore the Trust sued HYL for damages in negligence and was successful.

The important practical point here was the need to ensure that major areas of damage risk are properly insured, and that insurance is available in the case of disaster. In fact, the Trust was extremely lucky to be able to claim from a subcontractor that could be shown to have a duty of care, was still in existence and had adequate insurance cover. If the accident had been caused by the main contractor it would have been rather less fortunate.

The second situation is where the subcontractor has not caused any damage but has carried out work that is inadequate. This causes much more difficulty. The law of negligence is primarily designed to protect against physical loss or damage.

It is only in very special circumstances that the law has been prepared to allow damages for economic loss on its own, and to date the situation that has been accepted is that of negligent advice. Even then the claimant must be able to show that a special relationship of trust/dependence existed between the parties. The claimant has to have acted in reliance upon the advice of the defendant, who knew all along that his advice was being relied on and that the claimant would suffer loss if that advice was wrong and then the defendant must have been negligent.

Does a subcontract create a special relationship between subcontractor and client? This problem arose in the following case –

Junior Books v. *Veitchi Co* (1983) JB had a new printing works/warehouse built. V was a specialist subcontractor who laid a dust-free floor, comprising a plastic coating laid over a concrete base. V mixed the plastic wrongly and compounded the problem by not curing the material properly after it had been laid. The coating started to break up and needed to be re-laid. The cost of replacing the floor was £50,000, and the disruption to JB's operation was over three times as much. JB sued V for negligent work and was successful. The case went to the House of Lords and the court decided that the relationship between JB and V was sufficiently close to justify an economic loss claim. JB therefore recovered full damages.

At the time it was thought that this was to be a major extension of the law of negligence which at one stroke would solve the difficulties of cascading down liability in complex projects. But it was very quickly realised that the decision could cause major practical difficulties. If, for instance, every subcontractor could in theory be liable in damages every time any equipment or work supplied by him appeared to have failed within, say 15 years after a contract had been completed, and had caused economic loss to an end-user, then insurance costs, risks and prices would rise astronomically.

Even more importantly, it was also felt that the decision, even though made by the Lords, was not legally correct. It has never therefore been applied since, and in every case to which it might have applied the courts have 'distinguished the case on the facts' and refused to apply it.

A more typical example is the following case –

Simaan General Contracting Co v. *Pilkington Glass Ltd* (1988) S was the main contractor for a major public building being constructed in Abu Dhabi. The architect had specified that the main entrance hall/atrium of the building was to be glazed with panels of P's 'suncool 24/22 glass'. The glass was specifically designed for use in very hot climates and was also green in colour, green being the Muslim colour signifying peace. The glass was purchased from P, actually as a nominated sub-sub-supplier. When the glass was installed it was rejected for variations in colour. The coating had been wrongly applied so that the colour varied enormously, some panels even being red rather than green. S sued P directly in negligence but the claim failed. The court distinguished *Junior Books* on the basis that the facts were different, and the relationship was not therefore close enough to justify the claim. The correct route to follow should be through the contract chain.

In another case in 1986, *Muirhead* v. *Industrial Tank Specialities,* a case arising out of the failure of special oxygenating pumps installed in a factory fish farm for supplying fresh water to tanks containing lobsters, the result was the same. The contractual route should be followed.

13.4 The Contract (Rights of Third Parties) Act 1999

Although the Act creates major exceptions to the doctrine of privity it will not affect many contracts. It will only apply to a contract when the parties specifically agree that the Act is to apply, or allow it to apply.

The Act states that it creates additional rights for third parties but does not make any other changes to existing law.

It specifically does not apply to certain types of contract. They are self-explanatory:

* bills of exchange, promissory notes and other negotiable instruments;
* the agreement made between a company and its members by the articles of association;
* employment contracts and contracts with self-employed persons, home workers and others; and finally
* sea transport contracts and other transport contracts governed by international convention rules (except in relation to any limitations of liability included in the contract).

The Act deals with the rights of

* the 'promisor', the party who has given the contract promise; and
* the third party who wants to enforce that promise.

The position of the other party to the contract, the 'promisee', is not affected by the Act.

Section 1(1) of the Act states that a third party may enforce, in his own right, any term of the contract if –

- he is identified in the contract by name or is a member of a group; and either
- the contract expressly provides that he may do so; or
- the contract 'purports to confer' a benefit on him;
- unless the parties did not intend that term to be enforceable by him; and
- the benefit conferred by the contract will be the right to use a provision of the contract either to claim against the promisor, or to defend against a claim by the promisor.

Note the phrase 'purports to confer'. It is generally interpreted to mean that if a contract includes a clause or clauses that could be construed as giving a benefit to a third party, he will probably be able to enforce that benefit directly unless the contract contains a clause stating that the Act shall not apply, or that the benefit may not be enforced except by the parties to the contract.

The third party can be defined in different ways, but must be expressly identified in the contract. If no one is identified in the contract then the Act will not apply.

- The third party may be identified by name (this is quite easy to understand).
- He may be identified as a member of a class (tenants, subcontractors or grandchildren for instance).
- He may answer to a particular description ('any person using the premises', for instance).

The third party need not be in existence when the contract is entered into. This provision allows the promoters of a company to make a contract for the benefit of a future company, so that the company can then enforce the contract in its own name in due course, and future subcontractors can be covered.

The third party may use any remedy that would be available to him if he had been a party to the contract. This means two things. First of all it means that equitable remedies are allowed, so that in a situation such as that in *Beswick*'s case (an annuity), an order for specific performance would be available. Second, it confirms that in a situation such as *Jackson*'s case, the other members of Jackson's family would now be entitled to substantial damages.

A claim by a third party must take account of any claims already made and damages already paid between the promisor and promisee, and a claim by a third party cannot take the place of or cancel out claims by the promisee.

The promisor will have available to him all the rights of defence, set-off and counter-claims provided by the contract.

Finally, the third party will have all normal rights under the contract. He may sue. He is given a right to use arbitration if allowed by the contract. His rights are also protected. Once made for his benefit the contract cannot be varied or cancelled without his consent if it is foreseeable by the promisor that he has already relied or might wish to rely on it.

To date there has not been any significant litigation arising from the Act.

14

Other Relationships

14.1 Bailment

'Bailment' features in many commercial contracts. It creates an obligation of trust, separate from the contract. It happens when the owner or possessor of something, the 'bailor', puts it into the possession of another, the 'bailee', with instructions as to what to do with it. The bailee must take proper care of the thing and can now use it or do with it whatever he has been authorised to do by the bailor. Then when the bailment ends the bailee will deliver the thing back to the bailor or to someone else as instructed.

A simple bailment could be free, for instance, when we lend something to our next-door neighbour. He should now take reasonable care of it, the same care that he would take of his own property, and return it when he has finished using it. If he damages it he should repair it, or pay for it to be repaired.

Then bailment can be 'for consideration', under a contract. There are many types of bailment contract – contracts to repair, transport or pack goods, and so on. There are two distinct sets of obligations that the bailee then has. As a bailee, he has a number of duties. He will then have obligations under the contract.

In case of conflict the contract will prevail.

Bailment duties

The duty to keep the item safe
The duty is strict: the bailee will be liable for any damage or loss unless he can show that the loss or damage happened without his fault. A bailee should keep items properly insured.

Using Commercial Contracts: a practical guide for engineers and project managers, First Edition. David Wright.
© 2016 John Wiley & Sons, Ltd. Published 2016 by John Wiley & Sons, Ltd.

The duty to obey instructions

The bailee has a duty to comply with any instructions given to him by the bailor. Any failure to do so will automatically create liability. The clearest examples of this are cases where shipping companies have been held responsible to the owners of goods that they are carrying when the ship has diverted from the stated route, even where this diversion has not been the fault or responsibility of the shipping company.

Often contracts will limit or exclude the liability of the bailee. See Chapter 10 above. The approach of the courts to this in bailment cases is shown by two cases relating to the theft of valuable goods –

British Road Services Ltd v. *A V Crutchley & Co Ltd* (1968) BRS delivered a consignment of whiskey to AVC's warehouse at Liverpool docks. AVC was to store the consignment overnight and then deliver it to a ship. AVC's conditions of contract limited its liability for goods lost or damaged while in their possession 'for any reason whatsoever' to £800 per tonne. The whiskey was stolen. The theft took precisely eight minutes. Someone climbed a drainpipe to the roof, entered through a skylight, and used cutters to remove the padlock on the warehouse door. A driver then started a tractor that had already been hooked up to the trailer and drove it away complete with its load. There was no doubt that AVC was liable. The court was scathing about AVC's security. There was no permanent security guard or any burglar alarms at the site, despite the fact that AVC regularly stored high-value goods. AVC had even left the consignment in clear view to any passer-by for several hours during the day. However, since the conditions of contract were effective to limit AVC's liability, AVC were allowed to rely on them.

Morris v. *C W Martin and Sons Ltd* (1966) (See also Chapter 2.) M sent a valuable 'white mink' stole to B for cleaning. B subcontracted the cleaning to CWM, with the knowledge and approval of M. The contract between B and CWM was on standard conditions in the fur trade for this kind of work. They excluded CWM's liability for loss or damage to goods in its possession. While in CWM's possession the fur was stolen by the employee entrusted with the task of cleaning it, and was never recovered. The court held that as a 'sub-bailee' CWM had a duty to take care of the fur and not to allow it to be stolen. CWM was therefore liable as a bailee, and as a bailee was directly liable to M, even though there was no contract between them. (Remember that this case occurred in 1966, long before the Contracts (Rights of Third Parties) Act 1999; see Chapter 13.)

But since M had agreed to B sub-contracting the cleaning work, and B had subcontracted that work on the normal terms within the fur trade, M was bound by those terms, as this is what M had expected. But in a final twist the court then decided that on a strict interpretation, CWM's conditions excluded liability for loss or damage, but not for theft. Theft was neither 'damage' nor 'loss'. Therefore M was able to recover damages.

14.2 Security for payment – the mortgage, charge, lien (and the pledge)

The majority of the forms of security for payment used in commercial trans-actions are outside the scope of this book. However, a brief summary may be useful. Security may be given by a debtor in two ways: by using property as security – what is called 'real' security, or by the financial backing of another person as a surety or guarantor – what is called 'personal' security.

Real security can take four basic forms –

- a mortgage, usually used as security for long-term borrowing,
- a charge, often used as security for normal commercial borrowing (a bank will regularly require a 'floating' charge against the assets of a borrower as security for an overdraft or borrowing facility),
- a lien, to ensure the payment of a particular sum,
- a pledge (or pawn – the two terms are almost synonymous) used to borrow small sums on a personal basis.

Mortgages and charges are 'non-possessory' securities. The debtor remains in possession of the property. So to protect the rights of the creditor (if for instance the debtor should try to sell it) the security has to be registered. Secu-rities in respect of land are registered at the appropriate land registry. Securities in respect of the assets of a company must be registered with the Registrar of Companies. Only after registration, and certification of that fact by the registry, will the security be effective.

The lien is the right of a person who is in possession of goods under a contract to keep possession until he has been paid. Possession means either actual physical control of the goods or 'constructive' possession, the right to give instructions to a third party who does have possession. It can apply in a number of different situations –

- contracts to carry out work such as repair or modification on something belonging to another;
- contracts to supply services;
- contracts to supply goods to another – the unpaid seller's lien under the Sale of Goods Act;
- contracts to carry out work or prepare documents for another.

This list is not long, but it covers many areas of activity. Of course in nor-mal circumstances a lien will only be used very rarely. It is really a precaution, available for use if the seller/contractor is seriously concerned that he might not otherwise be paid.

Typical examples are –

- the right of a hotel to retain possession of any personal effects etc. left in its possession until its account has been paid;
- the right of a garage to retain possession of a car that has been repaired (but not a car that has merely been serviced);
- the right of a ship or carrier to retain possession of goods that they are transporting;
- the right of a manufacturer/seller (already referred to);
- the right of professional organisations to retain possession of documents that they have prepared.

The right will apply under any contract of the appropriate kind, unless excluded or limited by the terms of the contract. The right can only be used for as long as the holder retains actual or constructive possession. It ends as soon as possession is given up.

If the holder uses the lien he becomes a bailee, with the obligation to take care of the goods. The lien will end when the owner pays the debt or complies with the terms of the contract. If, of course, the owner fails to pay or perform, then the holder will have the right to give notice and then to sell the goods to recover his debt and to pay any surplus to the owner.

14.3 The pledge

The pledge is the transfer of possession of something from the owner to the pledgee. The pledgee will have the same rights as under a lien until the debt is repaid and the pledge redeemed.

14.4 Personal security

Personal security will normally take the form of a surety. A surety will be created by deed or by an ancillary contract to that between the creditor and the debtor, under which the surety will undertake to the creditor to meet the debt if the principal debtor defaults. It may take several forms. It may apply to a single transaction, or a group of transactions, or it may be continuing security over a period of time.

It may be either a guarantee of payment by the debtor, or an indemnity to the creditor against non-payment by the debtor. This distinction in particular has led to considerable amounts of highly contentious litigation. The result is that the whole subject has become something of a minefield, which is worthy of a detailed treatise on its own. Happily that is beyond the scope of this book. However, as an example of the problems ––

> *Yeoman Credit Ltd* v. *Latter* (1961) L bought a car on hire purchase from Y. L was a minor. Because of this Y required L to produce surety for payment. This was given by O, who signed Y's standard form surety agreement. L defaulted. Y claimed against O. The court decided that the hire-purchase contract was probably void under the terms of the Infants Relief Act 1874 (because a car was not a 'necessary'). Attention then switched to the surety contract. O claimed that the surety contract was a 'guarantee' not an 'indemnity'. This was critical.
>
> If the surety agreement was a guarantee that O would make any payments due from L, and the hire-purchase agreement was void, L did not owe anything so neither did O. But if the surety agreement was an indemnity to Y against L's failure to pay, then it would not matter whether the hire-purchase contract was void or not and O would be liable.
>
> The document signed by O was entitled 'Hire Purchase Indemnity and Undertaking', and the court decided that it was clearly an indemnity. Also, when it was signed, both Y and O knew that L was a minor. That was why Y wanted an indemnity. But the court commented that the law on suretyship in general was so complex that it was bringing the law into disrepute.
>
> Nothing much has changed.

14.5 Guarantees and bonds

A bond is simply a binding promise to pay a sum of money if called upon. A bond may take the form of a bond given by a finance company or a bank guarantee.

A finance company bond will be completed under seal or as a deed, or may be worded as a contract for nominal consideration. The bond will take the form of a promise to the creditor that if the debtor fails to pay or to perform, the finance company will compensate the creditor up to the value of the bond. Should the bond be called in, the finance company will then pay, usually having first checked that the claim is genuine. The finance company will usually require an indemnity from the debtor.

A bank guarantee is a 'standby documentary credit', usually issued in accordance with the International Chamber of Commerce 'Uniform Practice for Documentary Credits', the internationally accepted practice within the banking community. It will be issued usually as a deed and be subject to the law of the country where it is issued.

The law in England, like in most other legal systems, is that the giver of the guarantee should be free to comply with whatever has been promised in the guarantee. Some legal systems exert control, usually by insisting that the validity period of the guarantee must comply with local law, whatever the guarantee might say. (The basic intention is to prolong the life of the guarantee.)

It will be issued by the bank at the request/cost of one of its customers. It will be addressed to a third party and will be a promise by the bank to pay any sum up to a maximum if claimed by that third party during a stated validity period. The guarantee will then be given to the customer who will pass it to the third party in due course. Guarantees are often used as a form of surety, or as support for payments due under a particular contract. The guarantee will be irrevocable, in the sense that it may not be cancelled by the customer or bank without the approval of the third party. It may be unconditional, meaning that the bank will pay against any claim for payment by the third party, or conditional, meaning that the bank will pay against any claim made by the third party provided that any other conditions stated in the bank guarantee have been fulfilled.

In the event of a claim the bank will check that that it has been made in the proper form and that any condition stated in the guarantee has been complied with. If this is so the bank will then immediately pay the amount claimed to the third party, and then draw down the same amount from its customer. The bank will not concern itself with whether a claim is fair, or with any disagreement or dispute between the parties. It will simply pay.

Two examples –

Harbottle (Mercantile) Ltd v. *National Westminster Bank Ltd* (1978) H entered into three contracts for the supply of various goods, mostly coal, to a customer in Egypt. It was a term of the contracts that each contract should be backed by a bank guarantee for 5% of the contract price. There were problems. Under two of the contracts the customer complained about the quality of the coal, and he then refused to open the payment letter of credit required by the third contract. The customer then issued 'extend or we claim' demands in respect of all three guarantees. H extended them on a number of occasions. Finally, the customer claimed payment of all three guarantees. H obtained a temporary injunction to prevent NatWest paying against the guarantees. NatWest now asked for the injunction to be lifted so that it would be free to pay against the guarantees. It was clear that there were claims by the customer under at least one contract, so that H might have some liability to the customer. The court decided that the bank would be entitled to refuse payment under the guarantee only if it was in possession of clear evidence of fraud. This was not so. The court refused to interfere. The temporary injunction was cancelled so that the bank could pay.

United Trading Corporation S A v. *Allied Arab Bank Ltd* (1984) This dispute concerned five contracts between UTC and an Iraqi state agency. The contracts were for a total of over US$15 million and covered over 15,000 tonnes of meat and one and a third of a billion eggs. Each contract was supported by a 10% bank guarantee as a performance bond. The early contracts went well, but the final contract, for 1 billion eggs, was caught up in the disruption resulting from the start of the Iran/Iraq war. Delivery of the eggs was disrupted because the eggs could not be stored properly in Iraq. They were stored on a temporary basis but delivery was delayed. Three hundred million eggs went rotten before or immediately after delivery. Inevitably this caused a serious dispute between the parties. The customer insisted that all five performance bonds to be extended, and then claimed under them. UTC obtained an interim injunction to prevent AAB making payment against the guarantees; and AAB then brought the action for the withdrawal of the injunction.

The court took exactly the same view as in the *Harbottle* case. It would only intervene where the bank had clear information that claims were fraudulent. Again there was no clear evidence. Therefore the court withdrew the injunction so that AAB could make payment against the guarantees.

In its judgment the court said that when UTC took the contracts it must have been aware of the risk that a guarantee might be called where the claim was not properly justified. This was a purely commercial risk that UTC must be prepared to accept, and it should have priced for the risk.

This demonstrates the true nature of the risk. For a bond to be of any use it must be an irrevocable commitment to pay against a demand. If a demand is made, however unreasonably, then payment will be made, which the bank's customer will then have to meet – and in practice that is a risk which may be almost impossible to price, or insure adequately.

14.6 The 'letter of credit'

The letter of credit is also a documentary credit. It is used in many international trading contracts. It provides, as far as realistically possible, a guarantee of payment to the supplier, plus a guarantee to the customer that no money will actually be paid until the customer has control of goods that are in accordance with the requirements of the contract.

To understand how this is achieved consider Fig. 14.1 below. A seller (S) enters into a contract with a customer (C). The contract will cover the supply of goods by S and delivery to C. The contract will specify the goods and the shipping documents that will cover those goods when they are delivered.

If, for example, the goods are to be transported by sea, and to be delivered free on board a ship at a port in S's country for shipment to C, the shipping

Figure 14.1　The relationships under a letter of credit

documents will comprise a bill or bills of lading plus various certificates of conformity, packing lists and so on. The contract will also state that payment is to be made by means of a letter of credit to be opened by C at a bank in S's country.

C will then ask a bank usually in his own country, the instructing bank (IB) to open the appropriate letter of credit. (What C will actually do is to simply fill in IB's standard form.)

IB will agree the terms of the letter of credit and commissions etc. with an appropriate paying bank in S's country (PB). This will probably be done by an exchange of emails.

IB will then formally instruct PB to make the payments to S. This instruction will be given in the form of a document setting out the precise terms upon which PB is authorised to make payment, and for how long the letter will remain open/valid. This document is the letter of credit. PB will then notify S that the letter of credit has been opened, and usually supply S with a copy. The terms will specify the amounts that PB will pay and the documents that S must submit to claim payment, usually invoices, plus shipping documents, bills of lading and supporting documentation etc.

S will now deliver the goods and claim payment from PB. PB will check that S's documents are correct, and then immediately pay S if they are. 'Correct' means that the documents must comply exactly with what the letter of credit says, and the documents must be submitted during the validity of the letter. If the documents are incorrect PB will not pay. If PB does pay it will then draw down the same amount from IB, and send the documents to IB. IB will reimburse PB and give the shipping documents to C in return for payment by C. C can now use the shipping documents to collect the goods when they arrive at the port of destination.

This looks very complex but in reality is simple, because the procedure is so well organised. All that S has to do is agree an appropriate contract with C, including the details of the documentary credit and shipping documents that will support any claim for payment. All that C then has to do is fill in a form at IB, and the banks will do the rest.

The goods will then go from S to C. The shipping documents go from S to PB, to IB, to C. Payment goes from C to IB to PB to S.

But a word of caution. Under banking rules the paying bank must comply precisely with the instructions of the instructing bank in the letter of credit. Therefore the paying bank can only pay against a claim by the seller when that claim is made precisely in accordance with the terms of the letter of credit.

As an example –

J H Rayner & Co Ltd v. *Hambros Bank Ltd* (1943) R entered into a contract for the sale of groundnuts. Payment to R under the contract was to be by a letter of credit and a letter was duly opened stating that payment would be made against bills of lading covering 'Coromandel groundnuts'. R delivered the goods and claimed payment under the letter of credit. However, the bills of lading submitted with the claim described the goods as 'machine-shelled groundnut kernels'. The bank refused payment under the letter of credit because the bills were not in accordance with the terms of the letter. R sued the bank for non-payment.

The evidence was that within the trade everyone understood that Coromandel groundnuts and machine-shelled groundnut kernels were in fact one and the same thing. The court held that the bank was correct not to pay. The bank was required only to pay when the claim and the documents supporting the claim 'were in strict accord with the credit as opened'. There was no room for documents which were almost correct but not quite.

(In fact, the author has himself met with this problem. Many years ago there was a famous misprint in the US Armed Services Procurement Regulations. This was copied into the letter of credit for a contract which one of my companies had for the supply of delayed action fuses to the US navy to be used in demolishing a breakwater. The result was that we had to invoice for the supply of, not 'underwater', but 'underwear' demolition equipment. If we had not done so we would not have been paid.)

14.7 Documents of title

A document of title gives the holder the ownership and therefore the right to take possession of an item from another person who has possession or custody. There is one class of document that is recognised by English law as being for all practical purposes a document of title, and that is in the world of transport. It is accepted mercantile practice that only the holder of the shipping document, a bill of lading or an airwaybill has the right to claim possession of goods from the ship or airline when those goods have reached their destination.

14.8 'Letters of comfort' versus parent company guarantee

Letters of comfort

The problem is that letters of comfort may often be of very little practical value. The classic case is the following –

Kleinwort Benson Ltd v. *Malaysia Mining Corp Bhd* (1989) M, a company owned by MMC, traded in tin on the London Metal Exchange. To do this required considerably more capital than M had, even though MMC was a Malaysian government trading organisation. M therefore negotiated a loan facility with KB of £10 million. To begin with, KB asked that both MMC and M should be responsible for repayment of the loan on a joint and several basis, but finally KB accepted that MMC should merely provide support for M in the form of a letter of comfort.

MMC therefore wrote to KB as follows, 'It is our policy to ensure that the business of [M] is at all times in a position to meet its liabilities to you under the [loan facility] arrangements'. Later, the tin market collapsed at a time when M was indebted to KB for the full £10 million. MMC put M into liquidation. The debt was not paid by M and KB claimed from MMC under the letter of comfort. KB's claim in fact amounted to £12.26 million, the £10 million facility plus a further £2.26 million interest.

The court decided that the 'letter of comfort' was worthless. All it did was to state MMC's policy current at the time that the letter was written. It gave no guarantee that that policy would continue for the future, 'A statement of current fact is not a contractual promise, and was not intended as such'.

Parent company guarantee

The parent company guarantee contrasts with the letter of comfort. It is usually either a deed or an agreement for nominal consideration. It will be either a surety agreement, or a guarantee that if the subsidiary company fails to honour a contract, say because it has been put into liquidation, then the parent company will accept responsibility for ensuring that another company will do so.

15

Preliminary Agreements and Letters of Intent

15.1 Introduction

This chapter deals with two distinct but related situations.

Many purchasers, especially in the public sector, have to comply with complex approval formalities/procedures before they can place significant contracts. So time must elapse between completing the bidding process with a successful contractor and converting that agreement into a formal contract document. There is then a need for something, but not a 'contract', to bridge the gap between acceptance/work commencement, and contract.

The second situation is when a purchaser is part way through the process of reaching agreement with the successful contractor but needs work on the contract/project to commence even though the offer and acceptance process is still not yet complete. There is then a need to set up an arrangement which will allow the contractor to commence work on a secure payment basis while negotiations are concluded and the terms of the contract agreed.

The two situations are completely different. Unfortunately, the same device, the 'letter of intent' ('letter'), may be used for both. This results in significant confusion, as demonstrated in the cases.

15.2 Letters of intent

Letters of intent as bridging agreements

There is a series of cases in which the courts have had to deal with disputes arising out of this type of letter. For examples see *ERDC Group Ltd* v. *Brunel University* (2006) concerning a contract for the construction of a running track and associated sports facilities, or *Diamond Build Ltd* v. *Clapham Park Homes* (2008) concerning a contract for the refurbishment of housing units.

Using Commercial Contracts: a practical guide for engineers and project managers, First Edition. David Wright.
© 2016 John Wiley & Sons, Ltd. Published 2016 by John Wiley & Sons, Ltd.

In both cases the court held that as the terms for the contract had been agreed in full before the letters were issued, those agreed terms should be applied to the management and valuation of the work carried out. See also the analysis in *British Steel Corporation* v. *Cleveland Bridge & Engineering Co Ltd* referred to below.

In *ERDC Group Ltd* v. *Brunel University* the court valued variations, claims and counter-claims on the agreed contractual basis, completely setting aside arguments by ERDC that they should be valued on a *quantum meruit* basis. The court also commented that the reason why a letter was used was merely that both sides were too busy with other matters to get on with the work needed to complete the formal contract.

In *Diamond Build Ltd* v. *Clapham Park Homes* the court commented that the letter, counter-signed by the contractor, was nothing more than a simple contract, intended to remain in force until the formal contract had been sealed.

The true letter of intent

Properly used, a letter is a valuable tool in the contracting process. It fills a specific but limited need. At any moment there will be many letters in operation without any problems at all. Unfortunately they can occasionally be used incorrectly, and without proper understanding.

Complexity

The day-by-day needs of every organisation will be met by normal sales and procurement. These have the benefits of speed and efficiency. Contracts are placed on a reasonably standard basis for properly defined equipment and work on the basis of a clear contractual framework.

But it gets more difficult to place contracts quickly as requirements become larger and more complex, when they stop being a simple exercise in sale/procurement and become a project. There are two reasons for this. There is more to discuss and agree before the certainty necessary for a contract can be reached. Also more input may be needed from either side. The contractor may need to discuss with the purchaser what the purchaser's long-term as well as immediate requirements might be, what the real problems are that need to be anticipated, the demands of site or timescale, and what resources the purchaser may be prepared to contribute to the contract. The purchaser may need to discuss with the contractor what ideas and alternative solutions the contractor can offer. After all, they have very different skills. The purchaser will be expert at operating facilities, and will know what he wants those facilities to give him. The contractor will be expert at providing facilities and the problems that have to be overcome to provide them.

The result of all this is that the process of reaching agreement on the terms of the project cannot be treated as a simple matter of offer and acceptance. It goes beyond that. It may need in-depth discussions before the offer and acceptance procedure can even begin, and will almost certainly need more discussions after the formal procedure has been completed.

Time

Projects can come in many shapes and sizes. But what every project is concerned with is a starting position, a finishing result and a target date. And the worst project, to be avoided if at all possible, is the one with an unmoveable target date. (That is what makes a project concerned with an Olympic Games or World Cup so daunting.)

A project is always uncertain at the start. It cannot become a contract until the two sides have achieved certainty, both of the project and of the terms. Achieving certainty requires time. More often than not it will take longer, often significantly longer, to get through the pre-contract stages than the purchaser originally intended. In particular many contractual issues cannot be dealt with until after the technical detail has been agreed, and so have to be left until near the end. As a result the parties cannot always complete those contract negotiations before the work needs to commence. So to avoid delaying project completion beyond his target date, the purchaser will want to start the contractor working on the project while contract negotiations are still in progress.

The correct use of a letter is in this situation, when it is used as a short-term arrangement to allow the contractor to work while the parties reach final agreement, and then complete the contract.

Letters are not part of the law of contract. They are actually part of the law of what is now called 'restitution', and which used to be called rather inaccurately 'quasi-contract'. Restitution covers a number of situations in which one party has either –

- received a benefit as a result of something done by another in circumstances in which it was unreasonable for him to retain that benefit, or to do so without compensating the other party, or
- carried out work in anticipation of a possible contract at the request of the other in circumstances in which it is reasonable that the cost of that work should be borne by that other party.

Examples of the first type of situation are where goods have been delivered or money has been paid by one party to another by mistake, if say delivery or payment has been made to B instead of to C. In such a case it would be unjust to allow B to retain what he has received, and accordingly the law of restitution requires that money be repaid or goods returned (or equivalent compensation paid).

Letters come within the second type, situations where the law requires compensation for work that has been done. Compensation will be payable not on the basis of a price or pricing laid down by the parties, as in a contract, but on the basis of the reasonable value of the work that has actually been done, what is termed a *quantum meruit* – 'what it is worth' or 'what it deserves'.

The term 'letter of intent' is actually misleading.

First, the letter need not be a letter, but can be any form of written communication.

Second, a statement of intention by the (potential) purchaser to enter into a contract with the (potential) contractor is not by itself a letter. The letter must also ask, authorise or instruct the contractor to commence the work before that contract is actually put into place. The obligation to compensate the contractor can only arise where he has been authorised to incur expenditure beyond what would be normal in a pre-contract situation, and has then done so. The perfect example of this situation is the following case, both because of its facts and the legal analysis of those facts given by the court.

British Steel Corporation v. *Cleveland Bridge & Engineering Co Ltd* (1984) CBE had won a major subcontract to supply and install the structural steelwork for a building being built in Dammam in Saudi Arabia for the Saudi al Sabah Bank. For the steelwork CBE needed to buy 137 steel nodes, the connecting pieces required to join the different steel beams together. They had talked to BSC, and BSC had quoted a price of £209,000, based on preliminary designs and on using a new and relatively untried die-casting process. But, while BSC were prepared to quote a price, it was not prepared to accept the conditions of contract proposed by CBE, which were on stringent 'back-to-back' terms in respect of liability. However, time was passing, and even though there were several issues still to be discussed and agreed, CBE were very much attracted by BSC's price and needed work to start in order to meet the programme set out in the head contract. CBE therefore sent a letter to BSC stating that CBE intended to give BSC a contract for the supply of the nodes in due course, and requesting that BSC commence manufacture forthwith. The letter also referred to the price that had been quoted by BSC and CBE's conditions of subcontract as the basis for that subcontract. Despite its unwillingness to accept those terms, BSC commenced manufacture on the basis of the letter and continued the work until all the nodes had been delivered.

The actual text of the relevant parts of the letter is as follows:

We are pleased to advise you that it is [our] intention to enter into a Sub-contract with your company for the supply and delivery of the [equipment] …

The Price will be as quoted.

The form of contract to be entered into will be our standard form of sub-contract ... a copy of which is enclosed for your consideration.

We enclose specifications.

We request that you proceed immediately with the works pending the preparation and issue to you of the official form of sub-contract.

However, there were problems. The designs had to be changed. CBE produced a delivery schedule which proved virtually impossible for BSC to meet. So BSC was late. As a result CBE refused payment. One critical node in particular was seriously delayed by an industrial dispute within BSC. Finally, disagreement about the extent of liability that BSC was prepared to accept meant that no subcontract was signed.

Although the nodes were eventually delivered CBE was considerably delayed in installing the steelwork and received a claim for liquidated damages for lateness from its customer, the main contractor, for some £875,000. CBE claimed £666,000 from BSC (the difference between £875,000 and £209,000). BSC refused the claim and then sued CBE for payment, on the basis that the letter from CBE created the right for BSC to be paid on the basis of a *quantum meruit*. BSC claimed that there was no contract between the parties but that the letter sent by CBE created an obligation independent of the law of contract for CBE to pay. CBE counter-claimed that its letter, plus the commencement of work by BSC, in fact constituted a contract, and that CBE were entitled to claim damages for lateness from CBE.

Although the case did not progress beyond the High Court, which makes the decision no more than 'persuasive' in its effect, it was decided by a judge who was given rapid promotion to the Court of Appeal and then to the House of Lords, and was also joint author of the standard reference book on this area of law, which perhaps makes it more than a little persuasive.

The court said that there were three possible situations that might arise where the parties used a letter to start work before the formal contract was in place.

First, the parties might in fact simply create a normal bilateral contract, with the letter acting as an offer, the commencement of work as the acceptance, and with both parties then having obligations towards each other in the usual way. This was what CBE claimed the position to be.

Second, it might also create a unilateral contract, 'if you supply the equipment I will pay the price'.

But both these situations needed the parties to have agreed upon all the terms for the contract at the time of issue of the letter. (See *ERDC Group Ltd* v. *Brunel University* and *Diamond Build Ltd* v. *Clapham Park Homes* referred to above.)

But here there were major issues still to be agreed. Delivery dates were not yet agreed, the designs and therefore prices were not fully agreed, and the terms of contract, in particular the liability to be carried by BSC in the event of delay or defects for example, was not agreed either. Therefore it was impossible for the letter to form the basis of a normal contract, whether bilateral or unilateral. Nevertheless the letter gave clear authority to BSC to commence work on the manufacture of the equipment. This was the crucial point. By giving authority to BSC to carry out work which no reasonable supplier would expect to do without payment, when no contract could be in place, CBE had created a situation in which they became responsible under the law of restitution to pay for that work if and when it was carried out.

So BSC was entitled to be paid a reasonable amount for the equipment that it had supplied, based upon the cost of carrying out the work plus a reasonable profit. The actual costs incurred by BSC in supplying the nodes were found to be approximately £210,000. The judge therefore awarded BSC a sum of £231,000, that is cost plus 10% profit.

The judge also found that while the letter of intent had created an obligation for CBE to pay for the work that had been done by BSC, the law of restitution did not permit the letter to create any right for CBE to make counter-claims against BSC. The judge therefore dismissed the claims by CBE for the liquidated damages that they had had to pay to their main contractor.

The decision in *RTS Flexible Systems Ltd* v. *Molkerei Alois Müller GmbH & Co KG* (2010) confirmed the decision in the *British Steel* case but took the law further. The facts of the *RTS* case are not important. However, what happened was that after a letter had been issued the contractor commenced work and contract negotiations continued until substantial agreement upon the terms of the contract had been reached. At that point a major problem arose due to circumstances which were the responsibility, but not the fault, of the purchaser. This severely disrupted the work, and as a result, although the contract had been agreed, it was not signed. The decision of the court, the Supreme Court, was that at the moment that the parties agreed the contract the letter ceased to be effective and was replaced by a contract on the terms that had now been agreed – even though no contract was signed.

For the consequences of these two decisions see below.

15.3 Similar situations

There is also the situation that arises where exceptional work has been done by one party with the agreement of the other in the expectation that a contract would be completed. Two examples of this situation, both of which pre-date the *British Steel* case are:

Brewer St Investments v. *Barclays* (1954) BSI were the landlords of premises which B wished to lease. Discussions were proceeding normally and there was every expectation that the lease would be signed. At this point B requested BSI to carry out some alterations to the premises in anticipation of the lease. BSI did so. After the alterations had been carried out the parties fell out over the terms of the lease. (B wished to include an option to buy the premises. BSI were only prepared to offer first refusal if BSI decided to sell.) This proved a sticking point and negotiations collapsed. The court decided that the situation was governed by restitution. Neither side was at fault or to blame. They had quite simply failed to agree on a point that had not yet been agreed at the time the work was requested and carried out. But B had agreed to take the risk that no lease might be signed. BSI had been asked to do work by B which would not normally be done by a landlord before lease, and that work was clearly for the benefit of B, because it would have enabled him to make immediate use of the premises. It was only fair that B should reimburse BSI. The court pointed out that if after alterations had been carried out BSI had refused to grant a lease on any reasonable terms, then B would not have been liable to pay.

William Lacey v. *Davis* (1957) In the immediate post-war period the War Ministry operated a scheme to pay the cost of reconstructing buildings damaged during the war. D owned just such a building. In discussions with L, a builder, D led L to believe that D would give L the contract for reconstruction. D then asked L to prepare detailed estimates and calculations for the reconstruction work to enable D to apply to the Ministry for a grant. L carried out the work. D obtained the grant, then sold the building. The court decided that D should be required to compensate L for his work. The preparation of extensive plans/estimates/calculations in enough detail to enable negotiations to take place with the War Ministry went far beyond what a builder would normally be expected to provide before contract, and D benefitted from that work. Compensation was therefore awarded to L, and calculated on the basis of a *quantum meruit*.

Another case where the parties clearly do not make a contract but the law of restitution may imply an obligation to pay for work is typified by a 'fire brigade' case.

Upton-on-Severn Rural District Council v. *Powell* (1942) P suffered a fire at his farm. He telephoned the Upton fire brigade to deal with the fire, and the brigade did so. At the time both parties believed that the farm was within the Upton fire brigade area (so that P was entitled to the services of the fire brigade without charge). It later emerged that P's farm was actually in a different rural district. URDC therefore charged a fee for extinguishing the fire. The court found P liable to pay the charge, on the basis that although neither P nor the brigade intended to enter into a contract, 'the Law implied a contract' between the two. In *Lacey* v. *Davis* the court commented that this decision was questionable, but that P would in any event have been liable under the law of restitution. The work was done on a mutual misunderstanding that no contract was necessary for the work but that P accepted the risk that he might have to pay for the work at in the event that he had called out the wrong fire brigade.

When therefore can one claim payment for work under the rules of restitution?

- The claimant must have actually carried out work. If no work has been carried out, then nothing can be valued on the basis of a *quantum meruit* claim.
- The work must have been done after an event has taken place or a situation has arisen which will bring the law of restitution into effect. In the *British Steel* case the event was a letter. In *Lacey* v. *Davis* it was that the contractor had been led to believe that he would receive a contract. In the *Brewer Street Investments* case it was that there was an understanding that the premises were to be leased.
- The work must have been done as a consequence of that event/situation.
- The other party must have known that the work was being done and agreed to, or did not try to stop, the work being done.
- The work must be of actual or potential benefit to the other side, at the time it was commenced, even if it later turned out to be of little value.
- The work that has been done must go well beyond what would normally be done free of charge. The law of restitution is not there to allow people to recover the cost of work that they would do anyway. It only applies when people carry out work that is exceptional.
- In the event that the work was being done to further a possible contract between the parties, within which that work would be of value, it must be reasonable in the circumstances that the other side should take the risk of having to pay for that work if the contract fails to come into existence through no fault of either party.
- Finally, the actual costs which have been incurred in carrying out the work must be proved. Unless cost can be proved, cost plus profit cannot be recovered.

Before leaving this area some final comments should be made. Rights under the law of restitution are generally of only short-term duration, in that they bridge the gap between two parties until the completion of a contract. Once that contract is signed the respective rights of the parties will merge into and be governed by the terms of the contract. For this reason it is often the case that such rights are treated as of lesser importance than the rights that arise under a contract. This is particularly the case in respect of rights under a letter.

The practical problem is therefore ensuring proper control where work is carried out pre-contract. It is easy to ensure that people are disciplined in their approach to a contract. However, they may approach the pre-contract relationship with less care (see the *ERDC* case, where this did happen).

In respect of letters in particular the following should be noted, inter alia arising from the decisions in the *British Steel* and *RTS* cases.

- A letter cannot require the contractor to carry out work; he is entitled not to begin work if he does not wish to, or to cease work at any time.
- If the contractor does decide to carry out work he must comply with any conditions set out in the letter describing the work that he is to do or the quality required for that work, essentially specifications and scope of work. He will be bound by any limits to the quantity, value or duration of work stated in or referred to by the letter.
- However, the contractor will not be bound by any conditions of contract mentioned in the letter, in the sense that the purchaser will not be permitted to set up counter-claims against him based upon them, although the contractor could be required to comply with procedures set out in them.
- The purchaser can terminate the arrangement at any time before a contract is agreed, but will be responsible for paying for whatever work has actually been carried out by the contractor up to that point.
- The purchaser can gain a significant benefit from using a letter. Work can be started well before final agreement is reached, and the time pressure on the final negotiations is reduced. But the purchaser has to accept the risks of reduced bargaining power and disagreement arising after the letter has been used.

15.4 Preliminary agreements

It is often the case that at some stage during negotiations for significant or major agreements the parties will set out the outlines of the terms that they propose for that agreement in a preliminary formal or semi-formal document. Such documents are given titles such as 'Heads of Agreement', 'Memorandum of Agreement', or 'Preliminary Agreement' or 'Outline Agreement'. Of course, such documents have considerable commercial importance, but the question is whether or not they are legally binding.

There are only three rules that need to be remembered here –

- The first rule is that the title of the document is of no significance at all (see the case of *Branca* v. *Cobarro*, Chapter 5 above).
- The second rule is that the only thing that matters is what the document actually says. There cannot be a contract until the necessary terms have been agreed, and the parties have agreed to put those terms into effect. If the document does not do both of these then it is a mere 'agreement to agree'.
- The third rule is that an 'agreement to agree' is at best only an undertaking to make a reasonable effort to reach agreement on the terms of a future

contract. As such it is in fact almost inevitably unenforceable in practical terms.

If the document does contain all the terms necessary for a contract, without any qualifications to those terms, and then states that those terms should come into effect, then it may, and very probably will, constitute a valid contract in its own right. Again see *Branca* v. *Cobarro*.

16

How the Contract Ends

16.1 Introduction

The contract may end in one of four ways –

- performance – by being carried out;
- agreement – by the parties agreeing that the contract has been performed completely or to an acceptable extent;
- frustration – by its becoming impossible to carry out;
- breach – that allows the injured party to terminate the contract.

The vast majority of contracts end by being performed, or by agreement. A small number of contracts will end because of a breach. Very very occasionally a contract will end because it has been frustrated.

16.2 Performance and agreement

The law is simple. The contract must be performed. Anything less, or different, is not performance, even if it is of equal or greater use or value. Each party must carry out all his obligations under the contract completely and in accordance with its terms, unless the other party agrees or the failure is legally justified.

Simple contracts usually can be performed. The more complex the contract, and therefore the more complex the performance required, the more difficult it becomes to perform. So termination by agreement becomes more common.

Complex performance

In a simple contract, say for the supply of a car, to achieve complete performance is not difficult. The contract can specify make, model, engine capacity,

Using Commercial Contracts: a practical guide for engineers and project managers, First Edition.
David Wright.
© 2016 John Wiley & Sons, Ltd. Published 2016 by John Wiley & Sons, Ltd.

colour and so on. It is then easy for the supplier to obtain the correct car and deliver it to the buyer. The contract will also state how much the buyer has to pay to the supplier and when payment is due, so that the buyer can pay. So the contract can be carried out exactly, and be brought to an end by performance.

But exact performance becomes more difficult when the contract is complex or is of long duration.

There will be difficulty from the complexity of the goods or services being supplied. It is simple to supply a car. It is a different matter to construct an offshore oil platform or build a hospital totally in total accord with a specification which will run to several thousand pages. In the same way it is almost impossible to supply complex services to a client totally in accordance with the contract.

Then there are the problems. Personnel may move on. The business or regulatory regime may change. Particular services, software or components may become unavailable and have to be substituted. Currency or market fluctuations will intervene.

Finally, there are the mistakes. Human error will mean that some level of failure will always occur, probably on both sides. There will be delay, defects and variations.

The consequence is that most complex contracts are never performed to the letter, but both sides accept this. So the contract will be ended by an agreement between the parties once work has been more or less completed.

Entire obligation clauses

A contract may provide that one party only has to carry out a particular obligation under the contract (usually to pay) after the other party has completely performed one or more of his obligations. Typically the contract will provide for payment to the supplier of the total price in return for complete performance.

The typical example of an entire obligation clause of this kind in the commercial world is the payment clause often found in standard 'small-print' purchase conditions, which states that payment shall not become due from the purchaser to the supplier until the supplier has completed the supply of the goods or services being purchased totally in accordance with the contract.

As an example –

Cutter v. *Powell* (1795) C agreed to work as second mate on a ship on a long voyage. The contract contained all the normal conditions for a watch-keeping officer's contract, but the payment clause was non-standard. The normal terms for watch-keeping officers were for payment on a weekly or monthly basis, but C's contract with P, the captain, provided that he was to be paid a single fixed amount for the complete voyage. C died during the voyage (so that in fact the contract was frustrated – see below) and his widow sued P for payment for the part of the voyage leading up to C's death.

The court held that as the contract was on the basis of full payment for full performance, and as C had not fully performed, even though his failure to do so was not a breach of the contract, P did not have any obligation to pay. The contract had to be interpreted and applied according to the language it had used. As it said that a fixed sum was payable for a complete voyage, no payment was due unless the voyage was completed.

Entire obligation clauses are acceptable in simple contracts, but impractical in complex contracts. And the law will always try to reduce the scope of an entire obligation clause. In *Cutter* v. *Powell* the court was obviously unhappy about allowing the entire obligation clause to apply. In the case of *Hoenig* v. *Isaacs* (see below) the court said that when a contract is on the basis of a fixed price, but provides for progress payments, an entire obligation clause can apply only to the final payment/retention, not to the progress payments. And the court was also prepared to apply the doctrine of substantial performance.

The law provides a number of alternative solutions that have the effect of minimising the severity of the basic doctrine of complete performance.

The divisible/severable contract

A contract may be divisible or severable (the two terms are interchangeable) in two distinct situations.

Where it is for the supply of a number of separate consignments, or for a number of separate activities, depending on the terms of the contract payment may become due for each item as it is supplied or performed, and the purchaser will lose the right to reject that item if the rest of the contract is not performed. Of course this will not be possible if the individual items are so closely linked to one another that they are part of an indivisible whole. (See also section 31 of the Sale of Goods Act dealing with the delivery of non-compliant instalments of goods.)

Also, in contracts that result in the piece-meal supply of materials by the supplier to the purchaser, for example materials for carrying out repairs or work on site, payment may become due for those materials as they are supplied.

In leases and hire contracts, obligations may be uncoupled from each other so that they become independent. A typical example is the undertakings in a lease by the tenant to pay the rent and by the landlord to keep the property in good repair. So, the tenant will not be entitled to withhold rent because the landlord has refused to carry out repairs, and vice versa.

Part performance

Failure to perform the contract is breach for which the injured party is entitled to claim damages. But he must 'mitigate' his claim, that is, keep the amount claimed to a minimum. So those damages must represent only the loss actually caused by that failure and must take account of whatever benefit he has received from the part of the contract that has been performed. Therefore, although part performance is not complete performance it will affect the damages that can be claimed.

There are also two statutory exceptions to the doctrine of total performance that apply even where the contract contains an entire obligation clause.

Contracts for the sale/supply of goods. Under section 30 of the Sale of Goods Act –

1. Where the seller delivers to the buyer a quantity of goods less than he contracted to sell, the buyer may reject them, but if the buyer accepts the goods so delivered he must pay for them at the contract rate.
2. Where the seller delivers to the buyer a quantity of goods larger than he contracted to sell, the buyer may accept the goods included in the contract and reject the rest, or he may reject the whole.
3. Where the seller delivers to the buyer a quantity of goods larger than he contracted to sell and the buyer accepts the whole of the goods so delivered he must pay for them at the contract rate.
4. Where the seller delivers to the buyer the goods he contracted to sell mixed with goods of a different description not included in the contract, the buyer may accept the goods which are in accordance with the contract and reject the rest, or he may reject the whole.

The buyer must pay for whatever he decides to keep. If he decides to reject something that has been delivered, but which is usable, then he loses the right to claim damages (because he has failed to mitigate that part of his claim). To a large extent this avoids the unfairness to the supplier of the total performance doctrine (and any entire obligation clause) so far as they apply to any goods being sold or supplied under a contract.

Under the Apportionment Act of 1870, where –

• a periodic payment, such as rents, annuities or interest; or
• an income payment, such as salaries, wages or pensions;

is due under a contract, it will be deemed to become payable on a day-by-day basis. This exception prevents the doctrine applying to many regular payments.

Finally, if the injured party accepts all or any part of the work that the other party has completed, then he must pay a reasonable price for it. This principle is similar to that set out in section 30 of the Sale of Goods Act above, except that payment will be due not at the contract price but at a reasonable price.

Stopping performance

But these rules do not apply where a party has abandoned the contract –

> *Sumpter* v. *Hedges* (1898) S was a builder who had contracted to build two houses on land owned by H for a fixed price. (Compare *Cutter* v. *Powell*.) After carrying out about 60% of the work, and having been paid for part of it, S announced that he had run out of finance, and left the site, in effect abandoning the contract.
>
> H finished off the work, and to do so used building materials left on the site by S. S then claimed payment for the remainder of the work that he had completed. The court held that the contract was a price-based contract. For S to claim payment there had to be a new contract or H had to be in breach of the original contract. H had used S's material, but he could not leave the houses unfinished, because that would prevent him from making use of his own property. The work that had been done by S was far from complete, so that S had no claim under the contract, even for substantial completion. Therefore the claim by S for payment for his work failed.
>
> But the court did allow S to recover the value of the materials that H had used.

Substantial performance

This applies to the world of building contracts. It is a recognised exception to the rules of performance, but is an anomaly.

Where a contractor has 'substantially' completed the work required under a building contract he is entitled to claim payment of the contract price from the purchaser, less an amount to compensate for the work that has been left uncompleted.

Substantial performance has been heavily criticised. It is contrary to the rest of the law of performance. It is clearly directly contrary to the principle that only total performance of the contract can entitle the contractor to claim payment of the contract price, and that partial performance, except in the case of the exceptions referred to above, will only entitle the contractor to payment of a reasonable sum –

> *Hoenig* v. *Isaacs* (1952) H was a decorator/interior designer. He contracted with I to decorate and furnish a 'fashionable' flat in central London for a price of £750. The payment terms were 'net cash as work proceeds and balance on completion'. I paid £400 in three instalments and moved into the flat, but then refused to pay the remainder. H sued. I objected that the contract was for a lump sum price, even though there were progress payments, and that the work had been left unfinished. Therefore H was not entitled to payment of the full price.
>
> (I clearly considered that H's price was excessive for what he had done, and was simply trying to reduce the price.) However, the only items that were left unfinished were defects to a shelf in a bookcase and a faulty wardrobe door, together valued by the judge at only £56. The court decided without too much argument or

explanation of its reasons that as the work done by H was substantially complete, H was entitled to payment of the contract price, less the cost of repairing the two defects in the furniture.

When is work 'substantially' complete? That depends upon the contract. The work must have been carried out in accordance with the contract, with no significant items of work left undone and no major defects, and the work must be capable of being put to its intended use by the purchaser. But any shortcoming that renders the work unsafe, or not capable of being used in the normal way, will bar a claim of substantial completion. (In *Hoenig* v. *Isaacs* the key point was that I accepted that the flat could be lived in. He had also not objected to the defects until after he had moved in even though they were obvious.)

16.3 Termination clauses

Any contract may contain a term that allows a party to terminate. Complex or long-term contracts often contain clauses allowing one or both of the parties to terminate in the event of breach by the other, usually if the other party fails to remedy it after written notice of the breach. Contracts may also include a clause permitting termination 'for convenience' in other circumstances as well, with compensation.

There are some minor legislative limitations on the use of such clauses in relation to agreements for permanent or long-term property or financial arrangements, such as leases and credit/hire-purchase sales. However, the most important legislative limitation on clauses permitting unilateral termination is section 3 of the Unfair Contract Terms Act.

This bars the effectiveness of a termination clause in 'standard conditions' to exclude or avoid liability, except to the extent that is reasonable. For a discussion of the Unfair Contract Terms Act provision of reasonableness see Chapter 10 above.

Termination 'by agreement'

The law on the termination of contracts by agreement can appear very complicated. This is because any agreement to bring a contract to an end must itself either be a contract in its own right, or must be enforceable for another reason, usually by being made as a deed.

There are two quite separate situations.

The first, often called 'bilateral' termination, may happen when both parties have still to perform something under the contract. They can terminate by agreeing a settlement of all claims that they may have between them and then ending the contract on that basis.

The second, 'unilateral' termination, happens when one of the parties has completely performed his side of the contract. He may then confirm to the other party that he is satisfied with the position and therefore releases the other party from any remaining obligations under the contract. This should be done by deed.

The different ways in which this may be done are –

Accord and satisfaction

'Accord and satisfaction' is simply the legal jargon term used to mean ending an obligation under a contract by an agreement supported by consideration, effectively a bilateral termination.

Both parties will be reasonably satisfied with what the other has done, and both will recognise that they have not performed the contract perfectly. The result will be an agreement that the contract should be terminated on the basis of a final settlement of all the outstanding issues between the parties. This 'accord' will basically provide that each party will have no further claims against the other party under the contract (which provides consideration, 'satisfaction') on whatever terms have been agreed.

Commercially it is advisable for the termination agreement to be set out in writing.

Variation

What is meant here is not a 'variation' clause, which allows one of the parties to modify, change or extend the amount of work to be done by the other party under the contract, but a change to the contract.

Variation of the contract is not termination, but the considerations that apply are very similar. Any variation to the contract must be done by a separate contract in just the same way as an accord and satisfaction. As with an accord and satisfaction the form of the variation agreement is of no matter, except that in the case of the variation of any contract which is required to be in writing (see Chapter 5 above), the variation agreement must also be in writing.

Waiver and release

Waiver is the act of giving up a legal right.

Release is the legal jargon term that applies to unilateral termination. The party who has completely performed his side of the contract simply 'releases' the other party from his obligations to continue to perform the contract.

16.4 Discharge by frustration

A contract is frustrated where circumstances have intervened after the contract was made which are outside the control of the parties and have resulted in its becoming impossible to perform as originally intended. It is then terminated by operation of law, without any action by the parties.

It is much beloved of academic lawyers. It has led to a large amount of text-book law, plus an Act of parliament all to itself dealing with its consequences. It is also much used as an excuse for failure to perform, but the occasions on which the courts have accepted the excuse are few and far between.

Early law did not accept the possibility of frustration. The first case in which the courts actually agreed that a commercial contract might be brought to an end by being frustrated was –

> *Taylor* v. *Caldwell* (1863) T hired a theatre from C for four days to put on perfor-mances of a play. Before the performances were due the theatre was burnt down. With some reluctance the court accepted that in these circumstances the contract must come to an end since it could no longer be performed, and the basic pur-pose for which the contract was placed was no longer possible. So, as the rent was not due at the time of the fire it ceased to be payable.

The law only allows contracts to be frustrated where there has been such a radical change in the situation that the contract is impossible to carry out at all or can no longer be performed as originally intended. In practice this means that frustration is a matter of law. It depends upon legal interpretation of the terms of the contract for a decision on what the purpose of the contract is and whether it has indeed become impossible as a result of the event claimed as frustration.

> *Davis Contractors Ltd* v. *Fareham UDC* (1956) D entered into a contract in 1948 with F to build 78 council houses over a period of eight months for a price of £94,000. In fact, in post-war Britain so much reconstruction work was in progress that D could not recruit and keep an adequate workforce. The result was that the time required for D to complete the contract increased to 22 months and D's costs rose to £111,000. D claimed that the shortage of labour constituted a frustrating event. The claim was rejected by the court. A contract will not be frustrated just because circumstances make it much more onerous for one party or the other. And in fact, both parties had been well aware at the time the contract was made that finding adequate labour might be a problem, but this was a commercial risk accepted by D under the contract. (Compare this with the *Metropolitan Water Board* case; see below and Chapter 12.)

The *Davis Contractors* case is of great importance because it was appealed up to the House of Lords, which set out the basis of the modern law. The ques-tion is simply whether the contract is wide enough to apply to the new situation.

If the changed situation is still covered by the terms of the original contract then there is no frustration.

Inevitably, the categories of events that can cause frustration are fairly limited. Broadly they are –

The death or physical/mental disablement of one of the parties to the contract

This applies to contracts of employment or for services of a personal nature. An example is the case of *Condor* v. *Barron Knights* (1966) in which a contract between C, the group's drummer, and the pop group the Barron Knights was held to be frustrated. C's contract required him to be ready to perform whenever the group had an engagement. He became ill and was unable to perform more than a few nights a week. He was not totally incapacitated, but he could not comply with the essential element of the contract, to perform with the group in every performance.

Other causes of incapacity to perform, such as being sent to prison or being subject to bail or 'ASBO' restrictions, could have the same effect.

Supervening illegality

A change in law may make it impossible or illegal for one or both of the parties to carry out his side of the contract. This can happen in a number of different ways, so that it is worth setting out a few practical examples.

In any war situation, emergency legislation will make it illegal to trade with the enemy. All trading and other contracts with enemy nationals will then automatically be frustrated.

Also, during a war the government will often assume emergency powers to interfere with normal contractual relationships in the national interest. Equipment or materials may be rationed or requisitioned. Certain types of contract may be banned or severely curtailed; see for example *Metropolitan Water Board* v. *Dick, Kerr & Co Ltd* (1917) referred to in Chapter 12.

Changes in law of other kinds can also make contracts illegal and so frustrate them. For instance the introduction of exchange control regulations or import/export licensing requirements can render it impossible to carry out some aspect of a contract in a way which is not illegal, and if the contract allows no other way of carrying out that part of the contract then it must be frustrated.

Other types of legislation can also frustrate a contract. Examples include contracts affected by nationalisation or changes to employment law. In the same way the use of executive powers may also frustrate the contract, such as the operation of planning law or the use of compulsory purchase powers.

Destruction of the subject matter of the contract

An example of this is section 7 of the Sale of Goods Act, 'Where there is an agreement to sell specific goods and subsequently the goods, without any fault on the part of the seller or buyer, perish (or suffer radical deterioration or change) before the risk passes to the buyer, the agreement is avoided'.

The sole purpose becomes impossible

The easiest example of this is a 'coronation that wasn't' case –

> *Krell* v. *Henry* (1903) K advertised his apartment, which overlooked Pall Mall, for hire at a fee of £75 to view the coronation procession of Edward VII. H hired the apartment; the hire contract did not, however, refer to the fact that the hire was to view the procession. The king became seriously ill and the coronation had to be postponed. The court held that even though the contract did not mention that the room was hired to view the coronation procession it was quite clear that this was the sole purpose of the contract. Therefore when the procession did not take place this went to the root of the contract and the contract was frustrated. It was quite clear of course that the decision was a difficult one – because of course K was fully prepared to let H use the apartment on the day even though H could only have watched the traffic passing.

In contrast –

> *Herne Bay Steamboat Company* v. *Hutton* (1903) Following the coronation the King was scheduled to visit Spithead and review the fleet. (This was a very impressive spectacle with a large number of major warships, British and from overseas, and other craft assembled and 'on parade'.) Again, because of the King's illness the naval review was cancelled, but the fleet remained in position. H had hired the steamboat 'Cynthia' from HBS for £250 so that a large party could watch the naval review and then have a day's cruise round the assembled fleet. When the review was cancelled H refused to carry out the contract. HBS in fact was able to recover some of its losses by putting the Cynthia back into normal service, but then sued H for payment of the fee. The court held that because the fleet was still in position it would still have been possible to have a day's cruise around it. It was clear from the terms of the contract that its purpose was not only to observe the naval review by the sovereign but also to have a cruise. Therefore, as H had hired the ship partly for that purpose the contract had not been frustrated. H was therefore liable to HBS for payment of the remainder of the fee, plus the expenses that HBS had incurred in catering for the day and so on.

A supervening event

This is where an event occurs which prevents performance of the contract in the way foreseen by the parties. The example usually quoted is –

Jackson v. *Union Marine Insurance Co* (1874) In this case a ship was chartered in January to sail immediately from Liverpool to load cargo and then carry that cargo to San Francisco. The contract contained a normal clause that protected the owner from liability for delay due to the normal accidents of weather and navigation. The ship sailed but was then wrecked and was in dock being repaired for several months. The court held that the contract was frustrated, so that J was not liable to pay for the hire or repair of the wrecked ship. Even though the contract allowed for delay due to 'normal accidents', several months delay to a contract before it had even begun meant that the contract could no longer be performed in the way originally intended by the parties.

Compare *Jackson*'s case with the *Metropolitan Water Board* case, above. In both cases the effect of the event was to so disrupt the contract that the new situation was far outside the time/obligations envelope of the original contract, and so far outside that it created a totally different set of circumstances that were not contemplated by the original contract. Frustration of the contract may not take place in a number of situations:

Where the contract allows for an alternative method of performance

Lauritzen AS v. *Wijsmuller BV* (1989) This case concerned a contract to carry a drilling rig from Japan to Holland by either of two ocean-going barges, *The Super Servant One* or *The Super Servant Two*, owned by W. W planned to use the *SS Two*, and to use the *SS One* for other contracts. Before the contract could be performed the *SS Two* was sunk, through W's negligence. W then refused to use the *SS One* to transport the rig, claiming that the contract had been frustrated. L moved the rig by a more expensive method, and sued for the extra cost and delay. The court decided that although W was negligent in allowing the loss of the *SS Two*, it was not a breach of the contract. Nor did the sinking of the *SS Two* frustrate the contract, because it still allowed W to use the *SS One*.

Where an alternative way of performing the contract is still possible

For instance, when the Suez canal was blocked as a result of the Six-Day War in 1956, contracts to deliver goods that previously might normally have been carried via the canal and now had to be shipped via the Cape of Good Hope were not held to be frustrated, although they were much more onerous to perform. The goods could still be carried, although the ship had to take a different route. (On the other hand, a number of contracts relating to ships and cargos trapped in the Great Bitter Lakes in the centre of the canal were found to be frustrated.)

Where the party's own conduct has led to the contract being frustrated

Maritime National Fish Ltd v. *Ocean Trawlers Ltd* (1935) OT hired a trawler from M. M owned four other trawlers as well. M applied to the government of Canada for fishing licences for all five vessels. The government only granted licences for three vessels. M used the licences for three of his other trawlers, then claimed that the contract with OT had been frustrated, because OT could not use the trawler. The court held that because it was M's own decision that had made it impossible for OT to use the trawler, the hire contract had not been frustrated. The doctrine of frustration could only apply where a contract had been made impossible to carry out without the action of either party. So M was liable for breach.

The effect of frustration

The original position was that if a contract was frustrated it would end immediately, so that neither side would have any further obligation to the other. Any loss must therefore lie where it fell.

This could obviously produce a situation that might be unfair, and what is worse that unfairness might be a lottery. For instance, we already seen in the case of *Krell* v. *Henry* that H did not have to pay K the agreed charge for the hire of his apartment because payment was not due until the day of the coronation and the contract had been already frustrated before that date. In another case, *Chandler* v. *Webster*, the circumstances were almost identical, but the charge was to be paid in advance. The court held that the charge must be paid.

This was the position until the 1940s when the courts did accept that where frustration meant that there had been a total failure of consideration it was wrong to allow the loss to lie where it had fallen.

> *Fibrosa Sp Ak* v. *Fairburn Lawson Combe Barbour Ltd* (1943) This case concerned a contract made in June 1939, under which F bought equipment from FLCB for delivery to Gdynia in Poland and paid a 20% down payment. After the manufacture of the equipment had begun, war was declared. F claimed the repayment of the down payment, and the House of Lords accepted that *Chandler* v. *Webster* had been incorrectly decided and over-ruled it. But the decision was simply that as the contract had been frustrated, F should be entitled to repayment, because FLCB could not give any consideration by supplying the equipment.

Of course, this decision corrected the imbalance created by *Chandler* v. *Webster*, but just created an imbalance in the opposite direction. Now FLCB would not be paid for the work that they had done.

The situation was largely remedied by the Law Reform (Frustrated Contracts) Act 1943. The Act applies to contracts 'governed by English law', and deals solely with the consequences of frustration. Essentially it provides as follows:

1. all money paid or payable by one party to the other before the frustration event shall be repayable, or cease to be payable;

2. provided that the other party may be 'allowed by the court' to retain or recover enough out of those advance payments to cover all or part of the costs that he has incurred in performing the contract up to the time of the frustration event, should it be reasonable for him to do so;
3. all other payments due under the contract would then cease to be payable, and neither party would have any further liability to carry out any further work under the contract; and
4. finally, if either party has obtained a benefit under the contract before frustration, the court may order that party to pay to the other a reasonable sum in respect of the benefit, but not exceeding its value.

Note that the Act talks about the court's powers to order payment in respect of costs/expenses and benefits by the parties. Of course, the aim is that the parties will agree a reasonable solution for themselves.

In the very few cases that have occurred since then the courts have adopted a commonsense approach. Benefits must be valued at the time of frustration, and costs/expenses must be calculated on the normal commercial basis, with due allowance for overheads and so on.

So the position is now reasonably fair. If a contract is frustrated, costs and expenses can be charged against any down payments or progress payments received by the supplier, and each party must pay for any benefit that he has received.

But both parties still carry risk, the supplier that the contract will be frustrated at a time when he has not yet actually supplied goods or benefits and that his expenditure will have exceeded the amount of down payments and progress payments by the purchaser, and the purchaser that his expenditure in carrying out the contract will prove abortive.

16.5 Termination following breach

(See also Chapter 17 below.) Termination is a self-help remedy. Breach of contract does not bring the contract to an end – quite the contrary. The contract remains in force. But if one of the parties is in breach of a condition of the contract, or is in major breach of an innominate term, then the injured party is entitled to terminate the contract for that breach.

Termination for breach of a condition of the contract

A condition is a major term of the contract. So any failure to comply, however minor, is by definition a serious breach of the contract. If such a breach occurs the injured party has the right to terminate the contract, in addition to whatever other rights he may have.

It does not matter if the breach is only minor, or if it causes no loss. All that matters is that there has been a breach of a contract condition. See the *Union Eagle* case in Chapter 9 above.

Termination following a major breach of an innominate term

Termination for breach of an innominate term is only possible if the breach is significant. There is a real problem (see the *Kawasaki* case, Chapter 8 above) in deciding whether any particular breach is serious enough to justify termination. The consequences of getting the decision wrong can make it risky to terminate if there is any possibility that the other party will contest the decision.

The method of termination

The right to terminate will arise once the injured party is fully aware of the circumstances. At that point he has the right to allow the contract to continue, in legal terms 'to affirm the contract', or to waive his right to terminate. Any failure to terminate, in effect allowing the other party to continue with the contract without protest, or actively continuing with his own work under the contract, or requesting or requiring the other party to continue with the work, will have this effect.

Alternatively he may terminate the contract. The act of termination can either be a statement, preferably a formal written statement, made by the injured party to the other party to the effect that the contract has been terminated by reason of the breach, or any action by the injured party that will clearly demonstrate to the party in breach that the injured party considers himself no longer bound by the terms of the contract. In strict legal terms it is not necessary for the injured party to inform the other party of termination directly. Indirect communication of the fact of termination through a third party will be quite enough. In practice of course direct notification should be the norm.

17

Remedies for Breach of Contract and Defences to Claims

17.1 Introduction

Having a remedy is not the same thing as being able to use it.

The courts are very efficient at enforcing a legal right. If, for instance, someone owes money to another, the process to enforce that debt is straightforward.

But if, when someone claims a debt, the other claims that the money is not payable, or produces a counter-claim, the law is now dealing with a dispute. Now everything changes.

Instead of being able to claim immediate payment at low cost and within a few weeks or months at most, the claimant is faced with a long and probably expensive legal dispute. It will mean the employment for the duration of a team of solicitors and barristers, months spent identifying evidence and witnesses, preparing claim documents and then going to court. The time taken to get to court will certainly be in the region of several months, and may run into years. Depending upon the complexity of the issues, both of fact and contract interpretation/law, and the resulting amount of evidence that may need to be produced, the court proceedings may take anything from a day or two to several weeks. Witnesses will have to be produced in court when they are required, taking them away from other work, people will be occupied in talking to lawyers, explaining, briefing and assisting.

The costs will run into many thousands of pounds, fees for solicitors, fees for barristers and fees for expert witnesses, together with the cost of preparing and producing evidence. Even if he 'wins' the case and is awarded his 'costs', the claimant will still not recover more than about half the overall cost that he will actually have incurred.

What is worse, if at the end of all this the judge finds the evidence produced by the other side more compelling or the witnesses that they have produced more credible, the claimant may still lose, which will be very expensive both

Using Commercial Contracts: a practical guide for engineers and project managers, First Edition. David Wright.
© 2016 John Wiley & Sons, Ltd. Published 2016 by John Wiley & Sons, Ltd.

in terms of money and also perhaps in terms of reputation. Going to law is essentially a very high-risk strategy.

This is why the less costly and lengthy alternatives of arbitration and, especially in the construction industry, adjudication are more popular. But they are still expensive and high risk.

Recently there has been a big increase in the use of alternative dispute resolution (ADR) procedures, mainly mediation or conciliation. These are fast, and use guided negotiation techniques to reach a compromise. But they depend upon the willingness of both parties to admit that their performance may have been less than perfect. Where both sides are prepared to concede their mistakes this is a much lower risk strategy and avoids the need to litigate or arbitrate.

17.2 The decision to enter a dispute

Never go into a dispute, whether litigation, arbitration, adjudication or even ADR, without being very clear about what you intend to achieve and why. It might be to defend against an unreasonable claim or to recover compensation for significant financial loss. It might be to make a point or to protect reputation, or for another reason. But, whatever the reason may be, never lose sight of it. Disputes can develop a life of their own and continue long after they have lost any point and have simply become an expensive exercise in ego protection.

For just two examples of this among many referred to in this book, see the cases of *Aswan Engineering Establishment Co* v. *Lupdine*, Chapter 9 above, and *City and Westminster Properties (1934) Ltd* v. *Mudd* in Chapter 6.

In the *Aswan* case, once it became clear as a result of the expert witnesses' reports what the technical facts were, and that they were in agreement that the pails were more than adequate, it was very unlikely that the supplier was ever going to be found to be liable or even at fault.

Equally, in *Mudd*'s case, once it became clear what the commercial facts were, and what had been said between Mr Mudd and the agent for the landlord, it was equally clear that the landlord's case was very far from compelling, and that the court was going to look for a way to find in favour of the tenant.

In both of these cases the facts must have been obvious long before the disputes went to court. So why did the claimants waste time and money doing so when they might have abandoned the case or looked for a compromise? Maybe in *Mudd*'s case the landlord felt that it was still worth trying to make a point, but there really was no justification for persisting with what was an unwinnable case in *Aswan*.

Thankfully, disputes are comparatively rare. But to state the obvious: if a dispute does look likely the first thing to do is make a realistic assessment of your case. There are a number of questions that you must be able to answer, as a minimum.

- What are the facts – that you know and that you do not know, but that the other side may know;
- What have you done correctly – and incorrectly;
- What has the other side done correctly and incorrectly;
- Can you prove your claim;
- What have you said to the other side, and what have they said to you;
- Who said what, when and how;
- Have you said or done anything that might waive your rights or bar or stop (estop) you from claiming – see for example *Hartley v Hyams* (waiver) or *Central London Property Trust v High Trees House Ltd* (estoppel);
- Are the files of correspondence/emails/minutes of meetings/notes of telephone discussions and so on complete;
- If needed can you produce the people who agreed the contract and then managed it;
- Will they make good reliable witnesses;
- And, finally, does the amount in dispute justify the expense.

If the answers are not positive, then re-think. Simply, there may be little point in going into a dispute, or keeping on with a dispute, unless you have a reasonable chance of winning.

If the answers to all these questions are positive, there is still another question. If you have a good case, why does the other side still persist in thinking that it is worth his while continuing with the dispute? Of course the answer might be just that he is living in cloud cuckoo land, or does not want to give in too easily, or is just playing for time or hoping that you will blink first. Or he may be going into the dispute knowing that it will clarify issues in a way which will concentrate minds and lead to a compromise 'on the steps of the court'. Or he might just be too stubborn to admit that he is wrong. But the most likely reason is that he thinks that if there is a dispute, he will win. If he thinks that he will win the two most likely reasons are that he believes that he knows, and can prove, something that you do not know or have forgotten, or that he has interpreted the contract in a different way to you.

So there is one more two-part question that needs to be answered. What does the contract say AND what does the contract mean? Here of course there is always going to be a problem of interpretation. Everyone is to some extent at the mercy of their legal advisers, who may get the answer right, and may also get it wrong. There is no substitute for good legal advice. But even then do remember that a lawyer is like a computer. If it is properly programmed it will give the right answer. But if it is fed the wrong information and then asked the wrong question, the answer will be worse than useless. So prepare an impartial brief – explaining the commercial and technical background, the history of the contract and the dispute, your strong and weak points and the questions that you want answered, so that the lawyer can understand the situation before any discussions with you. Then select a lawyer who is expert, and experienced, in the field. Dispute is always a high-risk project. Get the best.

Perhaps the simplest example of getting something wrong through failing to ask the right question is the case of *Fellowes* v. *Fisher* referred to in Chapter 12. Solicitors wished to include a competition restraint clause in a clerk's employment contract. They therefore imposed a seven-mile restraint area, identical to the one that had been found to be reasonable for another solicitor's clerk in the earlier case of *Fitch* v. *Dewes*. But they did this without asking themselves why the court had decided that seven miles was reasonable. (It was only justifiable in a very rural area, but definitely not justifiable in the suburbs of a major city). Really a solicitor should have known better.

17.3 Breach in general

Law gives the injured party a range of remedies for breach, which he can use should he wish to do so. The reason why the injured party has several options is that breach can take many forms. Some are more serious than others.

If a contractor agrees to supply a thousand cast-iron widgets to a customer, it will be a breach if he then categorically refuses to manufacture or deliver them at all. It is also breach if he only delivers 990. It is breach if they are delivered several months late or two days late. It is breach if the quality of all or only two widgets is below that required by the contract. It is also breach if the widgets are not cast-iron, but 18-carat gold or platinum.

17.4 Remedies

Remedies fall into two basic types – self-help remedies and remedies that are available through the courts.

The self-help remedies include –

- the right to terminate the contract for breach of condition or serious breach of an innominate term;
- (including the right of a buyer to reject or refuse to pay for goods that do not comply with the contract); and
- the power/right of an unpaid contractor or seller in certain circumstances to exercise a lien or retain possession of the goods, and ultimately to re-sell them if the buyer is clearly unable or unwilling to pay for them.

The remedies available through the courts are: claims for damages for various kinds of breach, an action by the seller for payment of the price of goods supplied, claims for the equitable remedies of rescission and injunction and orders for specific performance.

17.5 Termination for breach of condition or innominate term, or repudiation

Termination is a right, not a duty. It is not possible for either party to force the other side to terminate a contract by his own breach or refusal to comply with its terms. The injured party always has the right not to terminate but to keep the contract alive in the hope or expectation that he will still receive some level of performance that will make the exercise worthwhile.

Termination is any communication or act by the injured party that tells the party in breach that the injured party has terminated the contract. Usually it will be by verbal or written notice by the injured party that the contract has been terminated, but other acts can also be notice, such as rejecting goods or withdrawing from or barring the other party from access to the place where work is being carried out. It is of course always advisable to give the party in breach notification in writing, or verbal notification later confirmed in writing, that the contract has been terminated. First, this will ensure that the actions of the injured party cannot be misinterpreted or claimed to be a breach of the contract in their turn. Second, it will fix the date or time of termination so that there can be no doubt later.

The right to terminate must be used properly or it may be lost, or at best suspended. Broadly speaking this means that it must be used or at least asserted as soon as the breach becomes apparent to the injured party. If it is not, then the injured party may be deemed to have waived his right in respect of that particular breach.

So when the breach does become apparent the injured party should either terminate or assert his right to terminate (for instance by telling the party in breach that he will, or may, terminate unless immediate action is taken to correct the breach).

If instead he acts in a way that could be seen as showing that he does not intend to terminate in respect of that particular breach, for instance, if he continues to press the other party to continue with work, then that right to terminate will be lost. Then he cannot terminate unless the other party continues in breach or commits another breach, when he can reassert the right to terminate by giving reasonable notice. (He will of course have the right to claim damages for the breach.)

For a statutory example of waiver see section 11(2) of the Sale of Goods Act, 'Where a contract of sale is subject to a condition to be fulfilled by the seller, the buyer may waive the condition, or may elect to treat the breach of the condition as a breach of warranty and not as a ground for treating the contract as repudiated'.

For examples of the suspension and reassertion of the right to terminate see the cases of *Rickards* v. *Oppenheim* and *Hartley* v. *Hyams*, Chapter 9 above.

And in the *Union Eagle* case, also Chapter 9 above, there is a clear example of the proper exercise of the power of termination for breach, in this case delay in performance. Immediately the breach occurred the party in breach was warned/informed by the other party that he reserved the right to terminate. Then when performance was tendered by the party in breach, the other party refused to accept it and formally terminated the contract. (It would have been equally proper if the injured party had immediately given notice of termination.)

17.6 Rights against the goods or other property

Liens in general

A lien (pronounced 'lean'), is the right of someone – agent, bailee or other person properly in possession of anything – to keep it (and ultimately to sell it if necessary) until his claims in respect of it, usually for payment of a debt, are satisfied. It is independent of the contract, and has developed as an alternative to the right to sue for payment, simply because the law recognises that in some circumstances possession really is nine points of the law, and legal action may very well be of little value, if for example the debtor is in serious financial trouble and therefore unable to pay. A typical example (of a bailee's lien) is the power of a garage to retain possession of a vehicle that has been repaired until payment of the bill by the customer, and similar liens exist in all other cases of bailment.

Rights under the Sale of Goods Act

The Sale of Goods Act, sections 38–43, creates two specific rights for the unpaid seller, a lien and a right of stoppage in transit. An unpaid seller is any seller who has not been paid in full (provided that he has not refused to accept payment when it is offered to him).

The right to enforce the lien continues only for as long as the goods are in the seller's physical possession, or that of his agents, after the sale has taken place. Once the goods leave that physical possession the right to a lien will cease, except where the seller has delivered them to a third party, such as a carrier, but reserved his rights to dispose of the goods. However, once those goods reach the buyer or any agent of the buyer, the right will cease.

The lien will usually apply in the following circumstances –

* when the sale was for immediate payment;
* when the sale was on credit, but the period of credit has expired; or
* when the buyer has become insolvent (that is unable to pay his debts in the normal course of business, whether actually bankrupt or in liquidation or not).

It can apply to part deliveries as well as deliveries of the whole of the goods.

Once the goods have left the possession of the seller en route to the buyer, the seller will still retain a right to stop the delivery of those goods to the buyer and to regain possession of them, if the buyer has become insolvent. Again this right will cease once the goods reach the possession of the buyer or any agent acting for him, or the goods have reached their destination and the buyer has been notified. The seller may exercise the right of stoppage by notifying the carrier/bailee in possession of the goods or by physically resuming possession.

Finally, the seller then has the right to re-sell any goods in respect of which he has exercised a lien or right of stoppage.

If the seller uses any of these rights, the original sale contract remains in force. In other words, if the unpaid seller can only re-sell the goods at a loss, he can still claim against the original buyer to recover his losses, if the original buyer has any assets that make the effort worthwhile.

Of course, these Sale of Goods Act rights and powers are of great importance, but their practical application is strictly limited. They hardly ever apply in consumer contracts and they do not apply to the vast majority of commercial contracts either. Their main application is in the world of international trade, when goods can be in transit or awaiting shipping instructions for weeks at a time.

17.7 'Retention of title' and 'vesting' clauses

Bankruptcy or liquidation of the other party is a risk for both buyers and sellers.

For the buyer the most common area of risk is in buying equipment. It results from timescale and progress payments. A buyer may well have to pay significant sums as down-payments or progress payments under a contract well before the equipment is ready for delivery. If the supplier becomes insolvent the buyer may lose the benefit of his advance payments. What may be even worse is that the buyer will have to restart the purchase of the equipment, resulting in his operations being hampered or delayed for months.

For the seller the risk is that he may deliver goods to a customer who then cannot pay.

Under the rules that apply to bankruptcy and liquidation, the trade buyer/supplier is normally an unsecured creditor ranking low in the pecking order of creditors, below government or secured creditors such as banks, and therefore generally loses very badly.

These considerations come into prominence during periods of financial stress, especially when an economy is in recession. They result in contract clauses that try to mitigate the risks by retaining or obtaining property in the goods.

Vesting clauses

For the buyer of equipment the answer is easy, the use of a 'vesting' clause. This will provide that once equipment is in manufacture or has been allocated to the contract, even if it is not yet in a deliverable state (and usually once the contractor has been paid at least part of the price of that equipment), it will 'vest in', i.e. become the property of, the buyer, although still in the possession of the seller.

The seller is now in the position of a bailee of the equipment. The clause will usually provide for the equipment to be marked as the purchaser's property, so that it can easily be identified if necessary. If then the seller does get into serious financial trouble so that, for example, he has to go into liquidation, the buyer can now claim the equipment as his property. Of course it is still in the possession of the liquidator, who has a lien until the buyer settles the bill for whatever work has been carried out on the equipment. But once that claim has been settled the buyer can take possession of the equipment, so that it can be finished or put into a usable state, possibly saving both time and money.

Retention of title clauses

For the seller the problem is much more difficult. After all he is selling goods to another. The whole purpose of selling goods is to transfer the property in those goods to the buyer so that the buyer can then use them. Once ownership of the goods has moved from the seller to the buyer it cannot be recovered except in very unusual circumstances.

The solution that emerged was the 'retention of title' or 'Romalpa' clause, again using bailment, but in the opposite direction, with the seller retaining the ownership of the goods after delivery until the buyer has paid the price, but allowing the buyer to use the goods in the normal course of business as a bailee.

(See section 19(1) of the Sale of Goods Act –

> Where there is a contract for the sale of specific goods, or where goods are sub-sequently appropriated to the contract, the seller may, by the terms of the contract or appropriation, reserve the right of disposal of the goods until certain conditions

are fulfilled; and in such a case, notwithstanding the delivery of the goods to the buyer … the property in the goods does not pass until the conditions imposed by the seller are fulfilled …)

But in this case there are many more difficulties to overcome. Suppose that a seller sells a picture to a buyer to hang on the wall of his office, on the basis that payment will be due some weeks after the picture has been delivered. The picture will stay a picture, and will stay easy to reclaim. If the seller inserts a clause in the contract that the picture will remain the seller's property until the buyer has paid the price that is perfectly acceptable. If the buyer does not pay the price then the seller will be entitled to take his picture back and sell it to someone else.

But suppose that instead of a picture the seller sells the paints that are going to be used to paint the picture. Once the picture has been painted the paints will cease to have any separate existence. They will have become part of the picture, along with the canvas and frame. What if instead of selling a picture the seller has sold the wallpaper? The wallpaper will also be hung on the wall but once on that wall it will have become part of the building and therefore lose any independent existence. It will in fact belong to the person who owns the building.

Then what happens if the buyer re-sells the picture before he has paid for it? The picture will now in effect have been turned into money, but can the seller now claim ownership of all or part of the money paid to the buyer for the picture in the same way as he could claim ownership of the picture? Money is not owned in the same way that goods can be owned. (For instance all banknotes are actually the property of the Bank of England.) All that any person can have is the right to be paid a certain amount, and in this case the seller would need to be able to establish a better right to be paid than everybody else.

The problem is that there are only three ways for the seller to establish such a right to be paid. The first is where that money has been paid to his buyer when acting as an agent, trustee or bailee for the seller, so that in effect the money was being paid to the seller himself. The second is where the money is paid into a bank account to which the seller has access. The third is where a 'charge' exists, when the seller can establish an equitable right to the payment which will be allowed to over-rule any other right in law. But for an equitable right to be enforceable against any third party, such as another creditor, the charge has to be formally registered, usually under the Companies Act. It is in practice almost impossible for a seller to create a charge without registration, and it is impossible in the normal way of things for trade sellers to be able to register charges on all their supply contracts.

(A typical example of the debt backed by a charge is the overdraft. When a bank provides an overdraft facility to one of its customers the facility will be backed by what is called a 'floating' charge registered against the assets of the company, so that if the company goes into liquidation the bank can seek to recover its overdraft by enforcing the charge against whatever assets the company has at the time. And the charge then takes priority over unsecured creditors.)

Finally, of course, the idea of retention of title is completely contrary to the established principles of bankruptcy/liquidation law.

With all these difficulties to surmount the retention of title clause has had a fairly chequered history and has achieved only limited success.

In the first case of its type, *Aluminium Industrie BV* v. *Romalpa Aluminium Ltd* (1976), a case concerning a large quantity of valuable aluminium foil, a retention of title clause which provided that the goods should remain the property of the seller until all sums owing by the buyer had been paid (a very wide clause) was upheld. The seller was therefore allowed to reclaim all unsold foil from the insolvent buyer. In addition he was also able to claim money paid to the buyer in respect of foil which the buyer had re-sold. The case was, however, quite unusual. The foil was re-sold without being altered and the payments for it were deposited into a separate bank account by the liquidator. It was also admitted by the liquidator/buyer that when he re-sold he was acting as a bailee for the seller.

In a later case, *Re Bond Worth* (1980), the experiment was tried of creating an equitable rather than a common law right of retention. The experiment failed because the court was only prepared to accept an equitable right after registration.

Then in two slightly later cases, *Borden (UK) Ltd* v. *Scottish Timber Products Ltd* (1981) (concerning resin used in the manufacture of chipboard) and *Re Peachdart Ltd* (1983) (concerning leather used in making handbags), the courts rejected claims under retention of title clauses on the ground that once the goods had been used in the manufacture of other products they ceased to exist as the original goods, even though still in the possession of the buyer. Therefore the seller's title must be extinguished at that time, and any further right must then be registered to be valid. The same view was taken by the court in *Clough Mill Ltd* v. *Martin* (1985) (concerning yarn for weaving into cloth).

But in another case, *Hendy Lennox Ltd* v. *Graham Puttick Ltd* (1984), where a Ford dealer had sold diesel engines to be installed into skid-mounted generating sets, the court held that as the engines were simply bolted into place on the skids and could be equally easily unbolted, they still retained their individual identity after installation and therefore were still the property of the seller. (This principle was fully endorsed by the court in the *Clough Mill* case.)

In the *Clough Mill* case the court raised another problem with retention of title clauses. Their purpose is to protect a seller who gives credit to his buyers for any part of the price against the insolvency of a buyer. But if the seller were to use the clause to recover possession of all the goods when, say, only a quarter of the price was still outstanding, he could then re-sell them, and get a considerable windfall at the expense of the other creditors. This is not permitted by the law, unless the clause then creates a trust for the seller to repay any excess profit to the other creditors (which would also need to be registered).

So what is the current position? It would appear that a properly drafted retention of title clause will be effective to retain ownership of the goods for the seller for as long as the goods remain in the possession of the buyer and still retain their original identity. Once that original identity is lost or the goods are sold to a third party, directly or by hire purchase, then the seller's title to the goods will end.

Once the seller's title has ended the only other right that the seller can have will be to follow the proceeds of any sale/disposal of the goods, or products incorporating the goods, by the buyer. Where the buyer has sold the goods as an agent or bailee for the seller, and treated the proceeds in such a way that they can be separately identified (see the *Romalpa* case), the seller can follow that money. Where there is no clear agency/bailment relationship created by the retention of title clause or the proceeds have simply been paid into the general funds of the buyer, then the seller will merely have the normal rights of any creditor unless he has registered a charge against the buyer.

Finally, the retention of title clause should be drafted to avoid the possibility that the seller could use it to make a windfall profit, perhaps by only permitting the seller to recover possession of as much of the goods as could be re-sold to recover his losses at that time.

However, the area is still a minefield –

In the case of *Caterpillar NI Ltd* v. *John Hall & Co (Liverpool) Ltd* (2014) CNI sold equipment to JHC for re-sale to a customer in Nigeria. There were problems with payment. The contract between CNI and JHC contained a Romalpa clause that was both complex and badly drafted. CNI claimed payment from JHC on the basis that the price was due and that property in the equipment had passed to JHC (and then from JHC to the Nigerian customer). The court held that the Romalpa clause managed to achieve the result that the equipment never actually belonged to JHC at all but that property passed direct from CNI to the Nigerian customer, bypassing JHC completely. Basically it retained ownership for CNI right up to the moment of re-sale by JHC, when it transferred direct to the end customer, without then creating any right for CNI to claim in respect of the price paid by the end customer. So CNI had no right to claim against JHC and the claim was dismissed. (Since JHC never owned the equipment it is doubtful that JHC could ever be liable, under the law of sale. Whether CNI could actually claim successfully against the Nigerian customer is also doubtful since there was no privity between them.)

17.8 Liquidated damages and specific charges

'Liquidated damages' is simply lawyers' jargon, meaning an amount agreed between the parties as a reasonable pre-estimate of possible loss and written into the contract. The damages then apply if a breach does happen, and will be paid in accordance with the contract. As such they are a self-help remedy, but are explained in more detail below.

Sometimes a contract may require a party to pay or reimburse another if he has caused the other to incur a specific cost or become liable to pay a charge. Again these payments will be made in accordance with the contract (but see below on the law concerning penalties).

17.9 Remedies available through the courts: damages

The theory is that in the commercial world damages are by themselves an adequate remedy for almost all breaches of contract. They should be sufficient to put the injured party into the position in which he would have been if the contract had been carried out correctly, including the cost of carrying out whatever work is necessary to put him into that position.

So the equitable remedies – an injunction or order for specific performance – will only be granted where it is clear that in the particular circumstances damages cannot be an adequate remedy, or that a further breach or breaches may occur.

But in practice of course this is not always the case. Damages can be a very inadequate remedy. When paid at the end of court proceedings that have taken many months or even years to complete they may be even less adequate. As a result the law accepts that especially in the commercial world it is normal for contracts to specify how damages for some of the more common breaches, such as minor lateness, disruption, or performance shortfalls of various kinds, might be calculated and then paid relatively quickly. So the courts are always reluctant to interfere in arrangements of this kind unless it is very clear that there is something badly wrong.

Damages are to be paid as compensation, not for profit. The injured party will only be permitted to recover the lowest level of damages consistent with recovering his loss.

There are two different basic ways of calculating loss: the profits that the injured party would have made if the contract had been carried out – 'expectation damages', or the abortive expenditure that the injured party has incurred which has now been wasted – 'reliance damages'. In theory they are separate, but in practice most claims will comprise elements of both.

Damages will be limited to those types of loss which the party in breach knew or should have known to be likely to occur and when that loss results directly from the breach. Damages will not be permitted in the case of losses that are too remote from the breach.

The types of damages

Damages may take various forms. The injured party may simply claim nominal damages, of £1 or £10 for instance, to signify that a right has been breached and that the other party is liable. This may happen in a test case or perhaps in a situation where the injured party is seeking a finding of liability on which to base a further claim in equity, perhaps for an injunction to prevent further breach. Nominal damages are always claimable for any breach of contract, even where the injured party cannot produce any evidence of loss.

The claim may be for a fixed amount, set by the contract or calculated as liquidated damages; see below. However, the normal commercial claim will be for substantial damages, damages that properly reflect the scale of loss suffered by the injured party.

Damages may be either general or special

Special damages may be claimed for the cost of any specific items, when the injured party is able to put a reasonably precise figure on the claim, such as the cost of repairing a damaged item or paying a third party to carry out the work left undone by the other. General damages will apply to areas of loss that cannot be quantified accurately. In claims for specific damages the injured party is asking the tribunal to award him all or part of a specific amount claimed in respect of a particular head of claim. In claims for general damages the injured party is simply asking the tribunal to assess the situation and then decide upon a reasonable amount based upon whatever evidence the parties can produce.

For instance, if there was a contract to dye cloth that was then to be sold, and a mistake in the dyeing process damaged the cloth so badly that it was totally unusable, a claim for the cost of replacing the cloth would be for special damages (so many metres at so much a metre), but a claim for loss of potential revenue or profit from the sale of the cloth would be for general damages.

The assessment of general damages may sometimes be a very difficult exercise, because the tribunal is being asked to value the unknowable. This does raise a serious point. The court will always do its best to decide upon a fair figure, but what a judge believes to be a fair figure after hearing evidence from both sides may be very different to what either party believed a fair figure might be before the assessment exercise began.

Claims for breach of contract and in negligence

It may sometimes be possible to claim damages under the law of negligence as well as under the law of contract if the act of breach is also actionable negligence. The two types of claim can overlap but it is important to distinguish between them.

In contract, liability is to compensate for the failure to complete the contract. In negligence, liability is to compensate for causing damage, and usually physical damage, of a kind that was foreseeable, plus any directly resulting economic loss.

In contract, all that is required is to prove breach without justification. In negligence, it is necessary to prove that there was also a failure to take reasonable care.

In one sense it is easier to claim damages under the law of contract, because the burden of proof is lower, but the potential extent of damages that might be claimed in negligence is wider.

(There are a number of other differences between the two. The time limitation periods can be different and will certainly run from different dates. Defences to claims, and the basis on which counter-claims may be made are different and so on.)

Claims in parallel

In a case in 1985, *Tai Hing Cotton Mill Ltd* v. *L C H Bank Ltd*, the court said that it should not be open in a case of breach of contract for the injured party to use negligence to claim more in damages than he could under the contract. This rule was applied in a number of cases where the court felt that negligence claims were being used to claim much higher sums as economic losses resulting from contract breach than would be permitted under the law of contract.

But in another case, *Henderson* v. *Merritt Syndicates Ltd*, only a year after the *Tai Hing Cotton Mill* case, the court accepted that parallel claims in contract and negligence were perfectly permissible where the facts justified it. The reason for this was that the duty not to cause foreseeable injury to another by negligence will not cease simply because there happens to be a contract between the parties.

(But in practice where parallel claims have succeeded the courts have been careful not to allow claimants to achieve significantly larger damages under negligence than would have been awarded under contract.)

So the situation is uncertain. Anyone contemplating parallel claims should tread carefully. The safest approach is to treat a claim for negligence, if it is available, as an alternative way of achieving reasonable compensation, not as a route to recover considerably more in damages than under contract.

Substantial damages

Substantial damages must give the injured party reasonable compensation for the losses that he has suffered as the result of the breach, but those losses must be of a type that is foreseeable by the reasonable person as likely to happen, and must be caused directly by the breach.

Where the contract concerns commercial dealings by commercial people they are expected to know what any person in the business would know, and therefore foresee the usual consequences of a breach.

> *Koufos* v. *Czarnikow* (1969) This case concerned a contract to ship a consignment of 3000 tonnes of sugar to Basra. Basra was a major market for sugar in the Middle East and it was well known 'in the sugar trade' that the market price would fall during December. The contract required the sugar to be shipped direct to Basra, with an expected arrival date of 22 November. In fact, the ship made two stops on the way and then made a further stop to refuel. So the sugar did not reach Basra until 2 December and the market price for sugar had already fallen. The owner claimed the loss in re-sale value of the sugar. The court held that the shipper should reasonably have foreseen that the price would fall in this particular market during the period of delay, so he was liable for the full loss claimed by the owner. The judges defined the basis of liability in different but very similar terms: that the party in breach 'should reasonably be able to contemplate the loss' or that 'there was a serious possibility' of loss, or that the loss was 'probable', or 'not unlikely'. And as the shipper was in the business of shipping sugar to Basra he was expected to understand how the trade operated.

There are two types of foreseeable loss, as explained in the following classic case –

> *Hadley* v. *Baxendale* (1854) H owned a mill. Production had to be stopped because the main driveshaft broke. H placed a contract with B to deliver the driveshaft to the manufacturer for repair. He did not tell B that the repair was urgent, so B did not hurry. There was therefore a significant delay to repairs to the driveshaft and H suffered a significant loss of production. He sued B for damages, claiming a large sum for his loss of profit/turnover. B's defence to the claim was simply that he was not aware that production from H's mill was at a standstill, and that if he had known that this was so he would have delivered the driveshaft much more quickly (and would probably have charged double the price). The court rejected H's claims.

In doing so the court defined two rules for when loss would be foreseeable.

Any person in breach will be expected to be able to anticipate the normal types of loss that might result from his breach, as the court put it, 'the losses that would arise in the normal course of things'. The commercial organisation is expected to anticipate all the normal commercial types of loss.

In addition to normally foreseeable types of loss the party in breach must also anticipate any further types of loss that the injured party had warned him about at or before the time that the contract was made.

Here we come face to face with the question of commercial risk versus commercial price. Often the price that the supplier will offer to a purchaser will to a large extent depend upon the amount of risk that he is asked to accept. Often the purchaser will be reluctant to disclose too much information about his commercial intentions or difficulties. The more risk the supplier is asked to accept, the more likely he is to include contingencies in his price and the more likely his cost is to increase as he concentrates more effort into performing the contract properly.

In *Koufos*'s case we see how the first rule is applied. The second rule was discussed and clarified in the following case.

> *Victoria Laundry (Windsor) Ltd* v. *Newman Industries Ltd* (1949) V was in the business of dyeing cloth. V's business was expanding. It had won some very lucrative contracts to dye cloth for the Ministry of Supply (actually for army uniforms). To cope with this V needed to increase its steam capacity and to do this it bought a second-hand boiler that was currently installed at N's premises. V emphasised to N that V wished to put the boiler into operation as soon as it could be installed, but did not tell N about its government dyeing contracts. The contract duration was three months. Due to a mistake by a sub-contractor to N who was dismantling the boiler it suffered significant damage and needed to be repaired. It was eventually delivered and installed some five months late. Because of the delay V had to resign its government contracts. V then claimed damages from N, both for five months loss of the expected extra profit from its laundry business, and also for loss of profit from its abortive government contracts. The court found that N was liable for damages for the loss of laundry business profit, but was not responsible for profit on the dyeing contracts, under the second rule in *Hadley* v. *Baxendale*.

The court defined the position in three simple statements. Where there is a breach of contract the injured party can recover the loss that has actually resulted to the extent that that loss is reasonably foreseeable as likely to result from that breach by the other party. What is foreseeable depends on what the party in breach actually knew at the time of contract. Knowledge is of two kinds, first what the party in breach ought to have known, as a normal person, and then what he actually knew in addition because of any other circumstances.

The heads of loss for which damages can be recovered

In the event of breach the injured party is entitled to recover enough compensation to restore him to the position in which he would have been if the contract had been performed. But he is not entitled to more than that. He is only entitled to the minimum amount necessary to achieve this. In legal terms he must 'mitigate' his claims; see below.

Loss or damage to physical assets
This is the easiest area to understand. If either party to the contract has caused physical loss or damage to anything that is the property of the other party by a breach, the cost of making good that loss or damage can be recovered as damages under the contract. And of course there would often also be grounds for a parallel claim in negligence.

The cost might be very high if for instance items of significant value, or premises, had been severely damaged or destroyed (see for example *Morris* v. *C W Martin and Sons Ltd* (1966)), perhaps considerably more than the value of the contract, but to recover that cost would be straightforward, provided that causation and contract terms are clear. Of course in practice damage could well be dealt with under the terms of insurance held by one party or the other.

Loss or damage to non-physical assets
Neither party will be liable to the other for substantial damages if breach causes damage to the reputation or goodwill of, or loss of business to, the other, though there would be liability for nominal damages. This type of loss is normally held to be too remote from the breach (unless it has been brought within the second rule in *Hadley* v. *Baxendale*).

However, there will be liability for damage caused to other intangible assets of the business of the injured party by a breach of contract. An example might be the misuse of confidential/proprietary information, copyright or software of the other party. Liability would be for any profit made by the use of the proprietary knowledge, and/or the loss of profit caused to the injured party.

17.10 Claims by the seller/supplier

Items supplied

Where goods or services have been supplied in accordance with the contract, the seller/supplier may claim the price for those goods or services as stated in or calculated in accordance with the contract, or a reasonable price under the Sale of Goods Act section 8(2).

The seller/supplier is not entitled to claim more than the contract price or a reasonable amount for goods or services supplied or carried out. If he has under-priced for the work, that must remain his problem.

Cost of work done

If a contract is terminated due to breach by the purchaser, the seller/supplier can claim the cost of the work that he has done, plus the costs of terminating the work, plus profit. (In this context 'profit' means net profit, not gross profit, for example, profit plus contribution to overheads).

In addition he may also claim any wasted 'reliance expenditure', that is cost incurred in reliance on the fact that the contract was in existence and could be expected to be carried out, even though it is not part of actually carrying out the work (such as the cost of hiring people or buying special tooling or equipment).

In one case, *Cullinane* v. *British 'Rema' Manufacturing Co Ltd* (1954), the Court of Appeal appeared to rule that the seller/supplier cannot claim both wasted reliance expenditure and profit as well, but must choose between one and the other. However, that case dealt with a situation where the claim make-up was unclear. There seems no logical reason why this should not be possible where the two claims are properly separated from each other and do not overlap.

17.11 Limits on claims

Mitigation of liability

Every claimant has an over-riding obligation to mitigate his claims.

If he chooses to replace a low-cost item of equipment that has been damaged with an equivalent but much more costly item then he is free to do so. But when he claims damages for the loss of the original item he must not claim more in damages that the minimum amount reasonably necessary to make good the actual loss caused to him. He must not expect to spend a large amount and then have an automatic right to recover that amount from the other side.

This raises a practical problem. If it becomes apparent to a supplier that the purchaser may be heading towards a breach of the contract, what should he do? Should he for instance seek to minimise his potential loss and the liability of the purchaser by stopping work and redeploying resources to other work or contracts? Or should he continue manufacturing at normal pace until breach occurs and the contract has to be closed down. The answer that the law would give is that he should continue at normal pace. Anything else risks breach by him of his own obligations. But in practical terms that simply increases his risk of a serious loss.

If a breach has occurred which may lead to termination, how quickly should the supplier look for an alternative customer or redeploy his resources? Should he do so immediately he has given notice to the purchaser that he is considering exercising his right to terminate, or should he delay in the hope of receiving a positive response? The problems are many and obvious and vary from industry to industry and contract to contract.

The answer that the law gives is that, provided that the seller/supplier has acted reasonably, he is entitled to take whatever action he thinks most likely to protect his business. He is after all an innocent party trying to overcome a problem caused by another. 'Acting reasonably' in this context will mean simply that he is entitled to do anything that could have been foreseen by the other party as a possible course of action in the circumstances.

Therefore, provided that the seller/supplier has adopted a reasonable course of action, and has then kept his claim arising from that course of action to the minimum, his claim would be accepted as reasonable, even if with the benefit of hindsight another course of action could have been less costly.

As a practical example, suppose that the supplier needs to replace a piece of equipment that has been damaged. How should he do it? Should he buy a replacement, or hire it? Should he try to find an exact equivalent, or can he hire something else that is readily available? What if the replacement will do the job that the damaged equipment was doing but at a higher cost? Again the same principle will apply – provided that the action taken by the injured party could have been foreseen by the other party as normal within the industry or as possible in the circumstances, the law will accept that as perfectly reasonable.

Limit to 'contract price'

The second limit on a claim by a seller/supplier is that claims for loss of profit, for work done in reliance on the contract and for the cost of work carried out under the contract will not be allowed to exceed the total contract price. (The only exception to this is when the buyer has caused damage to the assets of the supplier's business.)

The law is that it is not reasonable for any purchaser to be expected to foresee that if he defaults on a contract, that the losses he would cause would exceed the total payments which he would be due to make under the contract.

As has already been said, in cases where sellers have tried to recover higher amounts than this by using parallel claims for breach of contract and in negligence the courts have acted in practice to prevent claimants recovering sums that are higher than would have been deemed permissible under the law of contract alone.

Claims by purchasers and other parties

Loss of profit
The seller is entitled to claim the profit that he would have made from carrying out the contract.

The purchaser is entitled to claim the loss of profit that he would have made if the contract had been properly carried out, in other words, his loss of potential business as a result of the seller's failure to supply. (Of course the normal principle of mitigation of liability must still apply.)

This means inevitably that there is a considerable difference between the potential size of claim by the supplier and the potential size of an equivalent claim by the purchaser. Except in the case of damage or misuse of property, the supplier's claim is limited to the approximate value of the contract. The purchaser's claim may be many times contract value. Two of the cases quoted above, *Hadley* v. *Baxendale* and the *Victoria Laundries* case, were concerned with claims by the purchaser for damages of substantially more than the contract price in respect of loss or reduction of turnover for a significant period from a complete business unit as a result of delays or lateness by the supplier.

The legal principle is straightforward. If, when the contract is made, the supplier can foresee a loss as likely to occur if he fails to carry out his contract properly, then if that loss does result directly from breach by him he will be liable to compensate the purchaser for it.

Of course there are limits. The loss must be foreseeable by the supplier, under the rules in *Hadley* v. *Baxendale*. The purchaser must show normal commercial prudence in planning to keep the potential impact of the breach to a minimum. If the owner of a major production plant with a turnover of millions of pounds per week were to claim several weeks' loss of profit from the supplier of a small replacement part that was delivered late, when normal practice would be to hold replacement parts in stock, it is very unlikely that the court would allow the claim.

The purchaser would need to show that he would have been very likely to make the profit. Markets are not always predictable – they can reduce in profitability or diminish in size, and competitors may come and go.

Then he would need to demonstrate that the breach by the supplier was the sole cause of his failure to produce the products – in other words, that nothing else might have prevented or inhibited production.

Then he would need to show that he could not have sourced equivalent products from somewhere else.

Nevertheless such a claim can be very serious. Many projects in a range of industries are organised on the basis of a main contract supported by successive tiers of subcontracts, so that there is always the potential for a small subcontract to cause delay to a much larger project, whether that project is a production plant or a major office building, shopping centre, hospital and so on. The modern complex project structure increases the potential for major claims against main contractors and subcontractors.

Claims for reliance cost and financial loss

This is another aspect of the same general problem.

If a breach has directly caused financial loss to the other then the normal principles of all contractual liability apply. So long as the loss is foreseeable there will be liability.

The problem for the injured party is primarily that of demonstrating that the losses he has suffered are the direct result of the other party's breach of contract. One example of this is the case of *St Albans Borough Council v ICL Ltd*; see below. There was no doubt that the Council could demonstrate that it had suffered serious financial losses as a result of the failure of the computer program supplied to it by ICL, including loss of revenue/cash flow and government grants. The question was quite simply which of the losses were direct.

17.12 Liquidated damages

'Liquidated damages' simply means, in legal jargon, an amount agreed as damages between the parties (and specifically written into the contract) as opposed to the 'unliquidated' damages assessed by the court. Sometimes contracts will also refer to 'liquidated and ascertained damages'. This means liquidated damages agreed between the parties but fixed by reference to an external indicator, such as the rental cost of equivalent premises.

The principle is that when the parties to a contract know that a particular breach might happen, and that it is difficult at the time of contract to predict what the cost caused by that breach might be, they may agree a reasonable pre-estimate of the damage likely to be caused by that breach and then include in the contract a clause requiring the payment of that reasonable pre-estimate in the event that the breach takes place. Payment of that amount will then be in substitution for the normal rights that the injured party would have to terminate the contract or seek damages through court action.

These clauses raise a number of general issues that need explaining.

When they are used the breach is classed as a breach of warranty, not as a breach of condition. The contract may not be terminated, but the injured party will be paid the agreed damages however great or small the actual loss caused might be (and even if there is hardly any discernible loss at all). The contract will then continue.

This has a number of benefits. It maintains commercial relationships, and assists in better risk management. Because the damages are known almost immediately they can be paid immediately through the normal contract payment procedure. There is no need for lengthy investigations to ascertain costs caused by a breach, or for the parties to involve themselves in the expense and delays of a formal dispute.

Liquidated damages clauses are usually applied to predictable contract problems, such as minor lateness or equipment performance problems.

The issues that need to be examined are (1) the law as to 'penalty' clauses, what will constitute a reasonable pre-estimate of the loss that might be caused by a breach, and (2) what happens if the breach should extend beyond what is provided by the contract.

The law on penalty clauses

One of the classic quotations from judgments made by courts in the Victorian age was the statement 'the law of England knows not a penalty clause'. When the judge made this statement what he was doing was simply to put into the language of that time the broad principle of claims for damages. The purpose of the civil law remedy of damages, whether for breach of contract or negligence, is to compensate the injured party for loss. Damages are not for the purpose of penalising, or punishing, the party in breach. Punishment is a function of criminal law not of civil law.

What the judge was saying is that a clause that punishes the party in breach is not enforceable. Instead the injured party will only be permitted to recover the actual costs directly caused to him by the breach. So any clause in a contract that requires one of the parties to pay a specific amount in damages to the other in the event of a breach must pass one very basic test. If its purpose is punishment it will fail. If its purpose is compensation it will succeed.

Whether a clause is a penalty clause or a liquidated damages clause is a matter of law. Whatever the contract says, the tribunal is free to over-ride what the parties have said should it deem it proper to do so. In consumer contracts the tribunal will be much more prepared to find that a clause is a penalty clause than in commercial contracts.

The principles for determining whether or not a clause is a penalty clause were set out as long ago as 1915 in the case of *Dunlop Pneumatic Tyre Co Ltd* v. *New Garage and Motor Co Ltd*. They were set out by Lord Dunedin in the House of Lords and can be summarised as below.

- Terminology is not decisive. Whether the words used to describe the amount payable are 'penalty' or 'liquidated damages' is unimportant. The law will concern itself with the content of the clause not its language.
- A penalty is a sum that will frighten or punish the party that might be in breach – whereas liquidated damages are essentially a genuine pre-estimate of probable damage.
- The decision must depend on what could be foreseen by the parties at the date of contract, not on the facts of any actual breach.

- The clause will probably be a penalty clause if it stipulates a sum that is significantly higher than the greatest possible loss that could conceivably follow from the breach (because that will result in an illegal profit).
- Similarly, it will probably be a penalty clause if it stipulates that the same sum should be paid for a range of different breaches, some of which may be serious and some of which may be far less serious (because it cannot then be a genuine pre-estimate of loss).

In the *Dunlop Pneumatic Tyre* case the dispute concerned the validity of a clause in an agreement between the manufacturer and a wholesaler of motor tyres. The dealer had agreed that if it advertised or sold tyres at less than Dunlop's list prices, or if it sold tyres to anyone who had been 'suspended' by Dunlop, it would pay Dunlop a sum fixed at £5 per tyre. Covenants in restraint of trade of this kind were perfectly legal at that time. The court held that the amount of £5 was reasonable in the circumstances as a genuine assessment of the costs likely to be caused to Dunlop by a breach of its marketing policies. It also held that although the same amount was being applied to different types of breach, they were sufficiently similar in nature. The payment was therefore valid as liquidated damages. However, in another case in the same year, *Ford Motor Co* v. *Armstrong*, relating to a very similar distribution contract, the amount fixed by the contract as payment for breach was £250 per breach. (In 1915 this was more than the price of a house, let alone a car.) The court held that this was clearly a far higher figure than could be justified as a genuine estimate of costs, and the clause was therefore struck down.

A genuine pre-estimate of loss

In the *Dunlop Pneumatic Tyre* case Lord Dunedin said

> It is no obstacle to the sum stipulated being a genuine pre-estimate of damage that the consequences of the breach are such as to make precise pre-estimation almost an impossibility. On the contrary that is just the situation when it is probable that pre-estimated damage was the true bargain between the parties.

The problem is that in a complex commercial world, at the time a contract is signed it is almost impossible to predict what the situation will be weeks, months or even years later, and so what the consequences of an actual breach of contract might be. It might cause a major loss to the injured party but it might only cause a minor loss, and there could be a range of other possibilities between the two extremes. In those circumstances the parties may take a 'pessimistic' view of the possible cost, or an 'optimistic' view. Both views will be perfectly proper and both will represent a genuine but very different pre-estimate of the loss that might occur.

Of course, in practice the view, and therefore the figures fixed in the contract as liquidated damages, may well depend upon the practical bargaining power of the parties at the time the contract is made, but that is beside the point. If the figure that they have agreed falls within the range of loss that is possible from the most pessimistic to the most optimistic then that figure will represent a genuine pre-estimate and will be perfectly acceptable as liquidated damages.

The position when the breach is outside the provisions of the liquidated damages clause

The liquidated damages clause will apply to the specific areas of breach stated in the contract. All other breaches of contract will be governed by the general law. So in the event of any breach of contract outside the terms of the liquidated damages clause, the clause will simply not apply.

If the injured party wishes to revert to treating a further or excessive breach as a breach of condition, so that he may exercise his right to terminate the contract, it will probably be necessary to give the other party due notice that the breach must be remedied within a reasonable time, failing which he may terminate.

17.13 Equitable remedies

Historically, common law and equity were two completely separate systems. Common law came first and created the basis of most law – contract, tort, criminal law and so on. Equity then developed as an additional resource, with its own courts and judges, to improve on the common law system where needed, and to add additional rights and remedies where appropriate.

The two systems were finally merged into one late in the nineteenth century, so that any court can now apply the rules of both systems to any case.

The law is that in most situations where a breach of contract has occurred, especially a breach of a commercial contract, then termination of the contract, where permissible, plus damages will be an adequate remedy for the injured party. An 'adequate remedy' is one that puts the injured party into possession of enough money to make good any losses that he has suffered and to allow him to purchase similar goods or services from someone else.

Additional remedies will therefore only be allowed at the discretion of the court where the court is satisfied that in the particular circumstances of the case damages alone are far from an adequate remedy. (Note that it is only a court that can award an equitable remedy. They are not available through arbitration or adjudication, for instance.)

In addition any person who seeks an equitable remedy must be prepared to demonstrate that he qualifies for that remedy. The rules of equity are based upon the principles of fair or proper behaviour. Any person can claim damages, no matter how badly he has treated the other side, so long as he can show a legal right to those damages. However, no person will be granted an equitable remedy, however strong his claim, if it can be shown that he has acted unfairly or improperly towards the other side. If for instance he has behaved wrongly, say by withholding payments that may have been due without justification, then he may be barred by the court from an equitable remedy. 'He who comes seeking Equity must come with clean hands.'

The injunction

An injunction is an order by the court to a person to comply with a specific instruction given in the injunction. Injunctions are used in a number of different situations outside the field of contract, and interim injunctions are also regularly used as a part of the legal process, for instance, to prevent the loss of potential evidence. This type of injunction is often called a 'Mareva' injunction after the first case in which it was used, *Mareva Compania Naviera SA* v. *International Bulkcarriers SA* (1980), in that case to prevent a ship involved in litigation from leaving the UK.

As a contract remedy they are uncommon.

Very occasionally the court may issue a mandatory injunction, an order to a party to a contract to carry out a specific activity or item of work that he is required to do under the contract. This will usually only occur when there is no other practical way of getting that item performed, for example, requiring a landowner to remove obstructions to and then to maintain the proper flow of water in a stream or watercourse.

This is, however, fairly rare. The normal injunction will be a negative injunction, an order to the party not to do, or to cease from doing, something that he is doing, or not to repeat something which he has already done in breach of a contract. The common law position is that each individual breach of a contract gives the injured party the right to claim damages from the offending party, so that repeated breaches give rise to repeated claims for damages; however, in some circumstances this is clearly not an adequate remedy.

A typical situation where this might be so is where one party has given the other access to its information under the terms of a 'confidentiality' clause, and the other party has then misused that information. The initial breach is clearly actionable, as will be any continuing breaches, but a continuing series of legal actions over identical repeated breaches would clearly be impractical or impossible. In that case that the injured party will ask the court to issue an injunction to the offending party ordering him not to make any further use of the confidential information contrary to the terms of the contract.

The offending party can contest the request, and the court will not normally issue an injunction if it could be shown that it would be oppressive or unreason-able in some way. (However, where an injunction is sought because of a clear breach of a contractual obligation it would be rare for the court to give too much consideration to the inconvenience that it would cause to the party in breach.)

If then the offending party fails to obey the injunction, which is in fact a court order, he will simply be in contempt of court and subject to fine, and even possibly to the imprisonment of his directors or personnel for a period should the court deem it appropriate, until he has 'purged his contempt' and given undertakings in regard to future conduct.

The order for specific performance

An order for specific performance is the opposite of a negative injunction in practical terms. It is an order by the court to the party in breach to carry out the terms of the contract. Under English law the order for specific performance is very rarely granted. In practical terms the court will only issue an order when the injured party cannot, in practical terms, obtain equivalent services or goods from any person except the offending party. As a result an order for specific performance would never be granted to the commercial seller or supplier, and would only be granted to a purchaser or customer where the contract was for specific goods that were almost unique, such as perhaps a picture or statue, or where there was a significant element of personal skill or expertise involved in the production of the goods or services that were being purchased under the contract. (A practical example might be a contract for the purchase of unique software, where no other supplier in fact existed who could write equivalent software without incurring enormous development costs.)

Index